OLD & NEW

OLD & NEW

— Design Manual
for Revitalizing
Existing Buildings

FRANK PETER JÄGER [ED.]

BIRKHÄUSER
BASEL

OLD & NEW
DESIGN MANUAL FOR REVITALIZING
EXISTING BUILDINGS

Texts and interview Frank Peter Jäger
Coauthors Hubertus Adam, Anneke Bokern,
Winfried Brenne, Rainer Hempel, Claudia Hildner,
Franz Jaschke, Simone Jung, Julia Kirch,
Uta Pottgiesser, Susanne Rexroth,
Carsten Sauerbrei, Frank Vettel
Project management and editing Berit Liedtke,
Andrea Wiegelmann
Translation from German into English David Koralek/
ArchiTrans, Hartwin Busch (texts by S. Rexroth,
U. Pottgiesser and J. Kirch, R. Hempel, W. Brenne
and F. Jaschke)
Copy editing Richard Toovey
Art Direction, design and layout onlab,
Nicolas Bourquin, Thibaud Tissot, Eve Hübscher
Project management onlab Niloufar Tajeri
Cover design and concept onlab
Cover image Michel Bonvin
Plan graphics onlab Andrea Grippo
Illustrations Sam Green Illustration Studio
Fonts Korpus and Relevant Pro by Binnenland

A CIP catalogue record for this book is available from
the Library of Congress, Washington D.C., USA.

**Bibliographic information published by the German
National Library** The German National Library lists
this publication in the Deutsche Nationalbiblio-
grafie; detailed bibliographic data are available on
the Internet at http://dnb.d-nb.de.

This book is also available in a German language
edition (ISBN 978-3-0346-0523-6).

© 2010 Birkhäuser GmbH, Basel
P.O. Box, CH-4002 Basel, Switzerland

Printed on acid-free paper produced from chlorine-
free pulp. TCF ∞

Printed in Germany

ISBN 978-3-0346-0525-0

9 8 7 6 5 4 3 2 1

www.birkhauser.ch

LOCATIONS

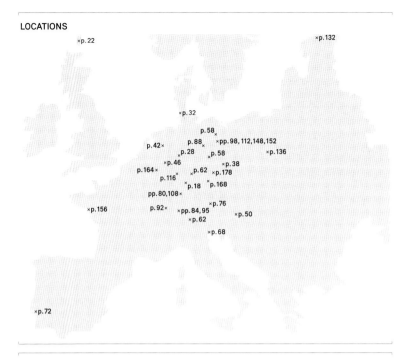

CONSTRUCTION COSTS
TOTAL 320 M €

ADDITION
17%

CONVERSION
41%

TRANSFORMATION
42%

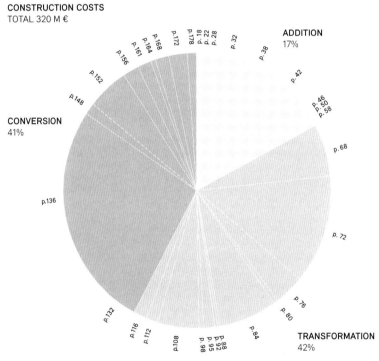

ADDITION

TRANSFORMATION

CONVERSION

IN DIALOGUE BETWEEN EPOCHS

Whereas, in the twentieth century, architects preferred the task of designing new buildings and only a handful of trailblazers had discovered the allure of interweaving historical and modern architecture, today the work of these pioneers has borne fruit; working with existing buildings has long since developed into an independent architectural genre.

FRANK PETER JÄGER

For a long time, there were three basic reasons for working with existing buildings: The first of these is that a new building is not affordable, or not seen as worth the investment, and so the old one is kept in use. The second applies when a building has landmark status and hence may not be demolished, but can instead be incorporated into a new development. In the third case, an existing structure is altered in lieu of demolishing it, because the attractive land use ratio of the existing building would no longer be permitted for a new one. Whereas the first reason is as old as the act of building itself, the other two are consequences of landmark preservation regulations and of modern planning law.

WHY WORK WITH EXISTING BUILDINGS?
—

In addition to the above, two motives pertaining to the ideal value of buildings play a role. Well before the establishment of modern landmark preservation, there were buildings considered to be sacrosanct—the birthplaces of famous personalities, sites associated with the acts of saints, and similar places with religious or political symbolism. The politics of symbols should be

> When buildings talk,
> it is never with a single voice.
> Buildings are choirs
> rather than soloists.
>
> Alain de Botton [1]

mentioned as a final, nearly forgotten, reason for working with existing buildings: In 690 A.D., when the building preceding today's church of St. Maria im Kapitol was erected in Cologne, the site selected was that of the masonry foundations of the Roman Capitolium, a temple dedicated to the divine trio Jupiter, Juno, and Minerva.[II] The site was, as we would say today, "newly occupied" as monotheism took the place of a polytheistic world throughout Europe.

Viewed in a historical perspective, our relationship to the existing built fabric is extremely dynamic: Whereas, for practical and economic reasons, working with existing buildings has generally been the rule during the two millennia since Classical Antiquity, it became increasingly seldom on the threshold of the Modern era in the nineteenth century until finally, in the twentieth century, demolition followed by new construction became almost universal. That has changed fundamentally in

1   Wall portal in Verona's Castelvecchio, which was converted to a museum by Carlo Scarpa

the meantime, above all in Central Europe, where approximately two-thirds of building activity now takes place in the existing fabric.[III] One reason for the pioneering role of the German-speaking countries—and some

neighboring countries—in this field lies in the comparatively early establishment of governmental landmark preservation in the second half of the twentieth century.

The monument culture of a local community cannot limit itself to restricting change and merely preserving a status quo if it seeks to maintain its influence amidst brisk construction activity. One of its most important tasks consists of providing architects and clients with reliable instruments for dealing with historically valuable buildings and ensembles: Should a converted industrial building be preserved in its entirety or just in certain parts? Is it permissible to blur the boundary between the existing and the new when carrying out alterations? In other words, should it be clear for all to see that the enlarged windows of a former warehouse building result from subsequent work? If yes, how can such points of intersection between old and new be formed? Lastly, there is the question of the authenticity of a building's various accretions: it is often decided to restore a building as a combination of several different historical states.

SMOKESTACK IN FREE FALL
—

Without discussing such questions, it is not possible to intervene responsibly in landmark buildings. Wherever the dialogue is conducted constructively and openly, architects do not interpret the conditions set by the landmark preservation authorities as a restriction, but as guidance for their design and planning. That the position of the landmark authorities—even on the basis of generally applicable fundamental principles—is not always explicable and is at times strongly affected by the subjective positions of individual decision makers, is heard in discussion amongst architects [p. 11]. On the other hand, several cases in which the architects have acted as advocates of buildings that did not even have landmark status are also discussed here. Every architect today needs to decide, independently from the assessment of the landmark

preservation authorities, what value to attach to an existing building for the purposes at hand—because the criteria of the landmark authorities considers the value of a building as historical evidence, but not all the possibilities it offers when considered from the perspective of an open-minded observer.

Ambitious combinations of old and new were indeed created in the 1950s and 1960s, yet compared with the current examples presented in this publication, the view of the past back then was defined by skepticism: in the reconstruction of Frankfurt's St. Paul's church, which had been bombed out in the war,[IV] or the Reichstag building in Berlin,[V] a mere few leftover exterior walls were celebrated as major feats of historic preservation, although in the cases of these and numerous

other buildings, much more of the built substance could have been saved. Moreover, such projects were almost exclusively limited to buildings of religious and symbolic significance. Disused industrial and transportation structures barely had a chance of being conserved until well into the 1980s. When I was a child, the newspapers frequently printed photo sequences of water towers being blown up, or factory smokestacks in free fall. Even then, I was not able to understand why there was supposedly no alternative to demolishing these fantastic industrial structures, as was always claimed.[VI] That such buildings, although originally constructed for thoroughly specific functions, such as water towers, silos, signal towers, and warehouses, could very well be converted for completely different uses is something that no one could—or wanted—to believe thirty years ago. The path to the preservation of these structures was blocked by a lack of architectural appreciation, not by their structural condition.

DIALOGUE BETWEEN OLD AND NEW
—

In many projects, the decision to conserve an existing building seems like an act of mercy by the architect to the ravaged residual fabric: Egon Eiermann's redesign of Berlin's Breitscheidplatz drew attention to the burnt-out tower that remained of the Kaiser Wilhelm Memorial Church, while all around, entire blocks of apartment buildings were being cleared away. Like the oversized spoils of war, buildings in Frankfurt, Cologne, and Munich were all torn from their contexts within an

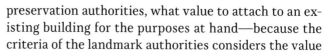

2   The breach torn by a bomb in Munich's Alte Pinakothek was repaired by Hans Döllgast in such a way that it remains legible.

3   In order to convert warehouse 11 on Cologne's waterfront into a residential and office building, JSWD Architects inserted additional windows next to the existing ones.

4   Restored portal on the front side of the warehouse in Cologne's Rheinauhafen harborside development

5   From water tower to residential tower: a project in Soest, the Netherlands, by Zecc Architects

ensemble, even though the neighboring buildings could have been rebuilt. Like a mere curiosity, the old pieces were isolated and elevated, but they were not conceived as equitable parts of a new whole. Many European architects first surmised what a true dialogue between old and new could look like when they saw the projects built by Venetian architect Carlo Scarpa in northern Italy. Since the late 1940s, Scarpa had attracted attention with a number of museum buildings, such as the renovation and conversion of the Castelvecchio in Verona and other architectural work on landmark buildings.[VII] His meticulousness in dealing with architectural details, his natural and equally elegant combinations of exposed concrete or delicate copper strips with the bricks, marble, terrazzo, oak beams, and other materials traditionally used in North Italian palazzi—and not least his pioneering, subtle way of dealing with the extant location—influenced a whole generation of architects. In Germany, Hans Döllgast[VIII] who was involved in the reconstruction of Munich, and somewhat later, Karljosef Schattner[IX] advocated a similar, pronounced position—although with completely different results. The work of these pioneers has borne fruit and has meanwhile found broad dissemination as an architectural stance: When architects and clients resolve to integrate parts of old buildings into new buildings, irrespective of landmark preservation requirements, they are doing so because they presume that the new entity will profit from the functional strengths, the presence, and the historical traces of such a building. What matters is the spirit of place and the historical period that a building represents, even if it has not made history itself.

The contemporary addition also enhances the existing substance. The act of inscribing another architectural epoch and the characteristic style of another architect in a building endows it with a new interpretation.

ARCHAEOLOGY ABOVE THE EARTH
—

The method of the Brückner brothers, which is first to have a look at a building at leisure and then conceptually consider whether a new building could replace its virtues [p. 12] speaks of respect for heritage and for their trained powers of observation. These stem from a sense for the barely perceptible qualities of a run-down, structurally reshaped building—in other words, from the ability to recognize and analyze concealed potential. In this situation, the work of architects resembles archaeology conducted above the earth's surface.

Nowhere do the social and sociopolitical aspects of working with existing buildings appear so explicitly as in residential buildings. Buildings from the boom years between 1960 and the mid-1970s especially —family homes as well as multi-story apartment buildings—with their clearly defined functional zones and the spartan dimensioning of service areas such as kitchens and bathrooms, are not reconcilable with the concept of living in open-plan layouts and with today's needs for space. By modifying the floor plan, adding stories, or through extension, buildings from this period can be upgraded in such a way that they receive a new, long-term perspective.

Every instance of the integration of existing structures in a new building, and every quali-

fying extension or renovation of an existing building equals practiced sustainability. The ecological significance lies first and foremost in the fact that an existent property that still easily fulfills its purpose continues to be used for as long as possible. Thus it is essentially a simple matter of thrifty and prudent economic management. In addition to the ecological benefit that is achieved by retrofitting a building for greater energy

efficiency, the savings in raw materials and energy that are attained by renovating existing buildings, as opposed to constructing new ones, should not be forgotten.

The successful combinations of old and new that are portrayed on the following pages should not obscure the fact that, internationally, the responsible treatment of building stock is far from being a well-established shared concept. In China, for example, which was economically opened up at the beginning of the 1990s, historical buildings have since been razed to the ground to an extent that takes away the breath of Western observers. In the Arab world, an awareness of the value of local building traditions has only recently begun to take root, as is the case in Central and South America. Even in North America, interesting examples of working with existing buildings are still rare.[X] Nevertheless, these are no grounds for Europeans to be complacent, because in Stuttgart or Vienna, for example, landmark status does not help a building much when its preservation is opposed by tangible economic interests.

Old & New is conceived in this context as an encouragement and an inspiration—for architects on both sides of the Atlantic—to gauge, on site, the opportunities for cleverly incorporating existing built structures, even if doing so may require them to be very persuasive. Proof that the results are worth the effort is provided by this

6   Historical traces in David Chipperfield's New Museum: a stair made of matte white exposed aggregate concrete between rough brickwork and pillars blackened with soot

7   Contemporary addition on top: a slender residential bar set on a house from the 1970s—a project by Wandel Höfer Lorch Architekten

8   Haus+ in Hameln: architect Anne Menke has reconstructed her parents' house in a contemporary manner.

selection of projects, which is heterogeneous in every respect; the majority is the work of small offices; many of them with very limited budgets.

THE PROJECTS
—

The chief criterion for selecting the projects in this book was their architectural quality, and thus also their originality. In this respect, a successful and multifaceted connection to "place" was one consideration, the conceptual maturity of the treatment of the existing fabric another. In light of these rather abstract criteria, it is surprising that in the end, a common theme can be discerned in the final project selection, without it having played a role in reaching a decision: A striking number of the buildings have a sociopolitical background; the new building is, in this case, an expression of the active re-cultivation of places that have in the meantime depreciated or become fallow. It is an expression of reurbanization. In the Dutch town of Enschede, the adaptive reuse of a factory site destroyed by an explosion a few years earlier has given the entire district a new identity [p. 42]. In Łódź, Tallinn, and Wildau in Brandenburg, the immense shells of buildings whose industrial use had become obsolete have once again, in a new guise, become part of daily life. They now serve educational, residential, or tourism purposes and likewise symbolize the newly found self-confidence of post-communist societies at the end of a transformation process lasting two decades. An outstanding example in every respect is Lesezeichen Salbke, because it demonstrates how the maximum effect can be achieved with minimum effort [p. 88].

The U-boat bunker constructed by the occupying German navy in Saint-Nazaire stood empty and useless at the harbor for decades: nothing other than a huge barrier that blocked the view toward the sea and constantly reminded people of the destruction of the town in the Second World War. This only changed when a Berlin architectural firm introduced cultural uses into part of the concrete

behemoth and in so doing, triggered the enhancement of the surrounding harbor district [p. 156]. The reanimated torso of the Kassel Zeughaus is a good example of the fact that sometimes even belated restoration projects, which have only been realized after several failed attempts, eventually achieve a particularly convincing result [p. 28]. The library in Schweinfurt [p. 172] is not the only design to convey the pleasure of gaining aesthetic capital from the well calculated confrontation between old and new. The affinity of such projects to modern art is obvious: Historical relics are inserted in new buildings like assemblages and there is a tangible enthusiasm for collage, deconstruction, and spatial layering.

On the other hand, the result need not always be a work of art. Those projects governed first and foremost by motives of practical use and endowed with limited funds admittedly appear less often in architectural magazines, but they make up the bulk of construction activity: extensions to industrial buildings, exhibition halls, schools and, not to be forgotten, the broad field of housing in existing buildings. The examples of Heuried residential complex in Zurich [p. 84] and Blumen Primary School in Berlin [p. 98] show that often a technically simple, but carefully implemented modification of the building envelope can give the whole a completely different character.

With two modern classics, the Stuttgart K II building [p. 108] and the former US consulate in Frankfurt [p. 116], the architects have consciously forgone setting their own accents and have concentrated entirely on refurbishing the technical equipment.

In the Franconian city of Würzburg, several hundred prefinished aluminum sheets, supported by a simple steel substructure, achieve a striking effect from a distance as the shiny envelope of a power plant [p. 18]. Working with existing buildings is not a discipline for those lacking in self-confidence. Those who believe that treating the original with respect prevents architects from creating designs with an urban presence in the existing fabric might well change their minds after seeing Würzburg cogeneration plant.

I   de Botton, Alain: *The Architecture of Happiness,* New York 2006, p. 208.
II   The Romanesque building visible today was completed in 1065.
III   In the last two decades, the percentages have continuously shifted toward working with existing buildings—for instance in Germany, new-built construction still dominated in 1997 with a share of 53.7% as compared to 46.3% for construction work in existing buildings, and the ratio has continually shifted since then: in 2007 in residential construction, for instance, the measures taken in existing buildings already comprised 74% of all construction work. Source: DIW (ed.): *Strukturdaten zur Produktion und Beschäftigung im Baugewerbe.* Berlin 2007. and: *Nachhaltiges Bauen im Bestand,* workshop documentation of the Federal Ministry of Education and Research, Berlin 2002.
IV   St. Paul's church was one of the first buildings in Frankfurt to be rebuilt after the Second World War, under the direction of the architect Rudolf Schwarz.
V   The architect of the reconstruction was Paul Baumgarten, born in Tilsit in 1900.
VI   Bernd and Hilla Becher, founders of the "Becher school," drove with their VW van through Europe in those days and took photos, from Wales to northern France, of water towers, blast furnaces, and mining pitheads, the majority of which have since disappeared.
VII   References incl. Los, Sergio: *Scarpa,* Cologne 2009.
VIII   Döllgast, who had mainly built for the church authorities since the 1920s, planned the reconstruction of numerous landmarks that had been damaged in the war, most notably in Munich. The "creative conservation" of the Alte Pinakothek, which had been damaged by a bomb hit, is amongst his best-known works. Cf. TU München (ed.): *Hans Döllgast 1891–1974,* Munich 1987.
IX   References incl. Zahner, Walter: *Bauherr Kirche—Der Architekt Karljosef Schattner,* Munich 2009; Pehnt, Wolfgang: *Karljosef Schattner—ein Architekt aus Eichstätt,* Ostfildern 1988/1999.
X   A much publicized project is, for example, the High Line Park in New York, an old elevated railroad track in Chelsea that has been converted by the architects Diller Scofidio + Renfro and landscape architects Field Operations into a city park. Further information under: www.thehighline.org

"A GIFT FROM THE PAST"

What makes working with existing buildings a challenge, how does it succeed economically, and what is its specific appeal? Claudia Meixner and Florian Schlüter (Meixner Schlüter Wendt Architekten) consider this in conversation with Peter and Christian Brückner (Brückner & Brückner Architekten).

INTERVIEW: FRANK PETER JÄGER

FRANK PETER JÄGER You are briefed to renovate and enlarge an existing building—what are your first steps?

CHRISTIAN BRÜCKNER You look at the building and its location, free of the impositions of the task at hand, in other words independently of the intended spatial program and the ideas of the client. You want to see what you will actually find there. We always seek traces, scratch away at the historical built substance, and explore the space. That is also our understanding of architecture: You can make a good contemporary design—but you cannot construct history. An appealing, multilayered existing building is like a gift from the past. It needs open-mindedness, fantasy, and sensitivity, in order to add something new in a convincing form. It is a manifold dialogue between old and new

that enables the new to emerge in togetherness, with each other or against each other.

FRANK PETER JÄGER What does this dialogue with the existing look like?

CHRISTIAN BRÜCKNER The most ecological position when building is, in fact, to use things that are already there; it can't get more sustainable than that. Moreover, a creative approach to the existing fabric is exciting and leads to enormous complexity in the course of the planning. In some cases, you resolve to intervene explicitly and extensively in an existing building, because otherwise no clarification—for example, no break from modifications made over time—would be possible. Often, structures have become superimposed over decades, and as a result, the legibility of the building has been changed entirely.

In the analytical phase, a clear view must be obtained: What has maintained value here and should be integrated—and what not? That is the purpose of this examination.

FLORIAN SCHLÜTER The crucial things for us are always the place we encounter and the task we've been given. It is good to let the surroundings, with their visible and invisible influences, have an effect. It could be that specific characteristics of the place have a much greater influence on our project than any building that may be there. Hence it would be superficial to consider an existing old building in isolation.   Thus, as a basic principle, we ask ourselves what influence comes from, for example, the topography, a large tree, or of course, an existing building. Ultimately, it is about making something that reacts to the place.

FRANK PETER JÄGER What criteria do you use to decide how comprehensive and drastic the intervention in the existing fabric will be?

CLAUDIA MEIXNER The criteria are different from case to case. Often, you cannot view the building detached from

its immediate context. The analysis of strengths and deficits would otherwise fall short. We have already been confronted several times with briefs to build in places that had lost meaningfulness. That is to say, the original character had been lost, because for a certain length of time there was no awareness of it. When we strengthen these places and revive their characteristics, we thereby also strengthen the overall urban structure. The decision on how intensely you must, and should, intervene depends on the character—on the impact—of a building within its context. The question becomes: What do you want to strengthen? Are there characteristics that it would be worthwhile to restore, and what remains of them?

PETER BRÜCKNER It is precisely this atmospheric facet, in the moment when you enter such properties for the first time, that is important. We like to take time to explore buildings that are to be renovated, to wander around their rooms. Preferably, we do it alone. Nothing should be a distraction. You allow the building to have its impact—and every old building and its individual spaces do something to whoever wanders through them. It is difficult to comprehend and even more difficult to formulate—a mixture of spatial perception, atmosphere, visible historical traces, evidence of past use, light, and

material. If we cease to feel that the interaction of these elements has a special power, we become cautious. We then consider whether something new—in the case of demolition—could be better than what already exists.

FRANK PETER JÄGER What about the argument of cost savings? Is working with existing buildings really less expensive than new construction?

CHRISTIAN BRÜCKNER It is problematic as a blanket argument, since much depends on the condition of the existing fabric and what needs to be integrated there. Working with existing buildings can at times also be more expensive than a new building. In our projects, however, including the existing fabric has always led to benefits for the clients.

CLAUDIA MEIXNER Maybe our reconstruction of the Dornbusch church in Frankfurt is a good case in point. At the outset, we considered demolishing the church completely and putting a small chapel in the courtyard as a replacement. Then it turned out that it would in fact be more beneficial to retain the sanctuary of the existing church. That was less expensive and also offered a potential for identification that was very important to the church community.

FRANK PETER JÄGER This ideal value of the building then became the reasoning for its preservation?

CLAUDIA MEIXNER Yes. At the time of its construction around 1960, the church was conceived for five hundred to six hundred people. In the meantime, the congregation on Sundays had dwindled to around thirty. Rain dripped through the ceiling, so there was a bucket next to the altar. When the church authorities announced that the church was to be demolished, the news was immediately spread by the press and the public was outraged. We then studied what could be done. Of the scenarios we investigated, partial retention was the best solution. During the discussions, it became increasingly apparent how important it was for the parishioners that we work with the existing building. Some of them had been baptized in that church. The building holds very many memories. Through our partial demolition, something like an accentuation resulted: The place and the memories of it were concentrated, consolidated within the former sanctuary. As part of the reconstruction, for example, we integrated a stained-glass window. Many people thought the window was new; but it came from the old church, where it

1   Dornbusch church in Frankfurt am   2   Public outdoor space and
Main after reconstruction: the former   "playing field" on the foundations
sanctuary is the new church interior.   of the demolished nave
The nave has been demolished.

had previously received little attention. In the new, smaller church, it suddenly dominated the space.

FLORIAN SCHLÜTER In the Dornbusch church, we have radically altered the existing situation. When you enter the church, you sense that something has happened there. But what is old and what is new cannot be differentiated with certainty; old and new are merged.

FRANK PETER JÄGER Some of your projects reveal an affinity with conceptual art...

FLORIAN SCHLÜTER Correct. With the Wohlfahrt-Laymann house, the old building stands like a Readymade within a new enclosure. It is, in fact, still very much identifiable—upon a second glance, at any rate. Much of the house is still there; we have intensified the presence of what still exists, by isolating it and placing it in a new context. Actually, it's a matter of the perception of things and their reflection—in other words the activation of an associative realm of experience. The house itself was of interest to us with a view to the superimposition of foreign elements.

FRANK PETER JÄGER In that project, too, the existing building was supposed to be demolished...

CLAUDIA MEIXNER Yes, the clients already had designs for a new building, but they were not really satisfied with them. Then we met and viewed the property together. The existing house was of solid, good quality, since it had been built by a carpenter for himself. We quickly developed the idea: Why shouldn't we work with this place, the house, and its history? The clients were not opposed to the idea, and were open to our proposals for integrating it.

FRANK PETER JÄGER What experiences have you had of landmark preservation authorities? Haven't these authorities played a significant role in the fact that conceptually ambitious projects involving existing buildings have meanwhile become commonplace in Central Europe?

PETER BRÜCKNER The landmark preservation authorities are a very important institution—especially when the dialogue and cooperation between architects and landmark preservationists leads to a good process: a planning

BRÜCKNER & BRÜCKNER ARCHITEKTEN,
TIRSCHENREUTH/WÜRZBURG
—

Klaus-Peter Brückner, born 1939, studied civil engineering at FH Regensburg
Peter Brückner, born 1962, studied architecture at TU München
Christian Brückner studied architecture at the Stuttgart State Academy of Art and Design.
Since 1990, the father and the elder son have run a joint practice; Christian Brückner, the younger son, joined the office in 1996. Peter and Christian Brückner hold guest professorships at the Munich University of Applied Sciences. Numerous awards and publications.

www.architektenbrueckner.de

process from which new stimuli for the design emerge. This is then reflected in the result. A good example is the Waldsassen cloister project, in which we have been working very productively with the landmark preservation authorities for ten years. You learn an unbelievable amount during design processes like that one. These stimuli often result in you jointly developing very different things than what was originally planned. We see that as a tremendous opportunity and it's also lots of fun. But of course, there are also misguided, dogmatic, landmark preservationists. In general, we dislike the term "landmark preservation"—a more appropriate term would be "landmark development," understood in the sense that people develop buildings and their surroundings further.

CHRISTIAN BRÜCKNER Yes, that is a crucial point. "Landmark preservation" often implies safeguarding and caring for a building. But then you can barely touch it or prepare it for a new function. As architects, however, we build living spaces. A landmarked building must live, or be resuscitated, when it stands empty—that's why I find the term "landmark development" much more contemporary. The other point is that "landmark preservation" does not exist. After years of collaboration with

many different such authorities, we have come to the conclusion that there are no binding, basic positions, in the sense of generally established positions on specific issues. So working with representatives of the landmark authorities depends on the person you are dealing with. In an extreme case, it then becomes arbitrary. The decision about matters such as how deeply I can intervene when making alterations to a landmarked building often depends strongly on the particular person involved. That cannot function as reliable orientation for clients and architects. We need more than just approvals from the landmark preservation authorities: It is, after all, also about jointly examining a historical building so that you can take it as a starting point and discuss its perspectives.

FLORIAN SCHLÜTER Ideally, a group effort emerges, and you jointly develop concepts for how a landmarked building can be placed in a new light. The goal, in our opinion, should be a landmark that can communicate its

3   Wohlfahrt-Laymann house near Frankfurt am Main: Meixner Schlüter Wendt Architekten have integrated the existing wooden house into their new building.

importance. It is not merely a matter of pure preservation, but also the comprehensible communication of various conceptual levels.

FRANK PETER JÄGER What significance does the subject of reversibility have for you?

PETER BRÜCKNER Reversibility is often a mandatory obligation to be followed in landmark preservation when intervening in built substance of historical value. In practice, this means that concrete slabs, for instance, may not be allowed; in other words, things that could only be removed with great effort are prohibited. We view this subject somewhat more subtly. Let's consider the Kultur- speicher in Würzburg: It was not, after all, built as a cultural venue, but as a warehouse. It demonstrates, however, how well this building type can also be used for other things. For me, that is an example of where it makes sense to establish new uses in a permanent and also irreversible manner. In making this decision, it helps to ask yourself: Is the perspective of reverting to the previous use one day realistic? Nonetheless, we naturally give thought to reversibility on a conceptual level. Especially in the case of particularly old buildings with valuable aspects, such as frescoes. These need not necessarily be restored. Often it is sufficient to preserve them effectively. That's very often the case with archaeological landmarks. Beneath a prudently designed new floor, historical floors that have been discovered during the archaeological survey can be preserved very well.

FRANK PETER JÄGER What role does the client's attitude toward the existing building play?

CLAUDIA MEIXNER Based on our experience, most clients have a relatively indecisive stance to their existing buildings, to begin with. In effect, we develop their stance together with them during the design process. We give them new views of things and influence them thoroughly. By analyzing what is encountered, the result is usually a clearly self-evident and resolute stance toward dealing with the existing situation. The spectrum thereby extends from the explicit preservation of what is encountered to its apparently complete disappearance.

MEIXNER SCHLÜTER WENDT
ARCHITEKTEN, FRANKFURT A. M.
—

Claudia Meixner, born 1964, studied architecture at TU Darmstadt and the University of Florence. Florian Schlüter, born 1959, studied architecture at TU Darmstadt and the University of Florence. Martin Wendt, born 1955, studied architecture at Frankfurt University of Applied Sciences. Since 1997, MEIXNER SCHLÜTER WENDT Architekten. Several awards, publications, and exhibitions, incl. Deutscher Architekturpreis, 2004; 1st prize, Wüstenrot Gestaltungspreis, 2006; category winner, World Architecture Festival in Barcelona, 2008; participation in Venice Biennale, 2004 and 2006

www.meixner-schlueter-wendt.de

CHRISTIAN BRÜCKNER We recently won a competition in which the brief was to raze most of a large, inner-city brewery and replace it with a new building to house a community center. As we examined the brewery buildings, it became clear to us that we must try to preserve parts of them. It's not just about a building. With the end of a use, some of the vitality of a place is lost. We then decided to take the role of mediator. We acted as the advocate for the existing fabric … and were lucky, because the municipal council recognized that more than one hundred years of urban history should not be thrown away without thinking twice. One aspect that should not be underestimated in such a situation, however, is economics: Linking the design of a community center to the existing buildings cost more in this case than a new building would have. But as a result, people get a spirited connection to history.

PETER BRÜCKNER The most frustrating experience for me in this regard was Maxhütte steelworks in Sulzbach-Rosenberg in the Upper Palatinate: a three-hundred-year-old smelting works. On our first visit, the plant was still in full operation, and we encountered a fascinating industrial landscape with equally fascinating spaces.   In this phase, we were allowed to develop conceptual models for where the journey could lead. In lengthy discussions, we considered how to save at least a few core elements of the plant, which had been a source of identity for decades. We also proved that it would be economically better to preserve many identity-forming elements and to leave them as a controlled landfill site. The site had significant contamination. We and our plans for the project were ultimately thwarted by, among other things, economic interests like those of the provider of waste disposal services. Firms in that field earn lots of money from the disposal of contaminated material. At some point we came to the conclusion that because it has similar cultural significance, the entire steelworks actually needed to be placed in the care of Bavaria's palaces and lakes authority. Ultimately, the plant was gutted piece by piece and sold. One smokestack is supposed to remain standing, but what's the point of that? A bitter experience.

FRANK PETER JÄGER Sensitizing industrial or utility companies to design issues appears to be difficult . . . the Würzburg cogeneration plant, however, is apparently an exception.

CHRISTIAN BRÜCKNER We thought at first, too, that we wouldn't find any common ground. The collaboration with the power plant's managers turned out to be surprisingly constructive; we really learned from each other. The objective of the preceding competition was actually

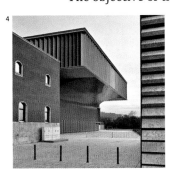

to build on top of the coal bins on the harbor side of the power plant and to consider possibilities for a new building. Our approach then evolved rather quickly in the direction of conceiving the entire existing power plant anew. The starting point for the expansion was to increase efficiency and optimize the operational procedures. At the beginning, there was great skepticism on the part of the client and the contractor consortium about us architects; architects do not ordinarily have equal footing when working on such projects. Nevertheless, it was decided that everything pertaining to the public image of the building would become our task, and we took part in the planning meetings. When it came to the position of the smokestack on the expanded power plant, we asked: "Couldn't there be three chimneys instead of the two that are planned, and couldn't they be arranged differently?" There were quizzical, astonished glances all around. But they thought about it and came to the conclusion that it would be worthwhile to consider different approaches in order to realize the best solution. What followed was mutual gauging of the situation, from which increasing trust then developed, as well as attention to the things that mattered to us.

FRANK PETER JÄGER What, for instance?

PETER BRÜCKNER One example is the so-called "nose"—the arched cantilever on the east side of the cogeneration plant. In the meeting where the client representatives were to make decisions about the exterior form, we naturally got the question: Can the nose be left off? It costs money and is not functionally necessary. We replied: Take a look in the mirror and then ask the question again.

FRANK PETER JÄGER What was the reaction?

PETER BRÜCKNER The nose was built. We naturally had to demonstrate that we could sheath the power plant's 10,000 sq m outer skin as designed for 2.2 million euro. To clad this huge area on schedule and within the budget was an audacious endeavor in terms of planning. We had promised the client that we could execute the project for

the set budget. That alone creates great pressure to succeed. The project was on a knife-edge more than once.

FRANK PETER JÄGER When dealing with existing conditions, how does one learn to distinguish healthy compromises from rotten ones?

FLORIAN SCHLÜTER There are certainly cases in which you have failed to some extent—subtly failed at a more-or-less high level. That has to be said clearly. When you are working with a budget ceiling, but after beginning design work, more cost-intensive solutions emerge as the best, there are basically just two possibilities: completely stop the project—something no one wants—or somehow implement it more economically. There is, in other words, a good idea, but whether it can be realized with the given budget is questionable. What you ultimately produce still represents a respectable design service. But you have a tear in the eye, because you know it could also have been much better.

CHRISTIAN BRÜCKNER It's important that the architectural concept can withstand things. It must be robust enough to withstand, for instance, cutbacks in the materials, or minor changes in the planning. At the same time, you have to keep an eye on the basic idea—otherwise the project suddenly no longer corresponds to what you originally wanted. The power plant in Würzburg was expanded again, two years after completion of the facade, which meant yet another smokestack. That would have the destroyed the symmetry of the three smokestacks. The solution was ultimately to insert the fourth flue in one of the three existing smokestacks. Interestingly, this also turned out to be the most cost effective solution. This persistent thinking outside the box, which must be cultivated whenever technicians and designers collaborate, is indeed often strenuous and time-consuming, but it is also enriching.

FRANK PETER JÄGER Keyword cost-effectiveness: In small-scale projects requiring intensive coordination, isn't there the risk that you slip into the red financially?

CHRISTIAN BRÜCKNER That happens. Regrettably, the fee structure for architects does not constitute a useful benchmark for actual expenditure on some projects. There will always be projects with insufficient fees. These alone would not support the office, but they can be carried by other projects. On some projects, it's not possible to do anything but invest in a lot of detail work. Beyond the monetary reward, there are other, sometimes more important factors, such as the satisfaction of the people for whom we create living environments and the recognition that comes with that.

4   The revamped Würzburg cogeneration plant features a cantilevered "nose," which shelters the new harborside terrace.

ADDITION

Extending, adding stories, enlarging, integrating, supplementing, rounding off, enclosing—the multitude of terms for the construction measures that can be summarized under "addition" immediately reveals the broad range of design possibilities. Their aim is to gain more space, upgrade a building to enable new uses, or even—when a new enclosure is placed around the existing building—enhance the appearance. In this respect, "adding-on" sounds slightly crude, because the expression conveys nothing about the opportunities for qualitative gain. Successful projects are ones in which the existing and new buildings merge to create a unity; suddenly you can no longer imagine the old without that which has been added.

In such cases, the architecturally attractive old building informs the newly added part, lending it aesthetic maturity and ambience. The coexistence of widely disparate building phases contrasts attitudes and layers of time; it generates unexpected spatial sequences, rough transitions, etc.—many things that an entirely new building cannot offer. The old and the new strengthen and enhance each other, and the contrast inscribed by the architecture inspires the users. The new whole is more than the sum of its parts.

WÜRZBURG CO-GENERATION PLANT

WÜRZBURG (DE)
Industrial — **pp. 132, 136, 148, 152, 161**

— Brückner & Brückner Architekten

1   Like one of Sant'Elia's dreams: the renovated Würzburg cogeneration plant, seen from the harbor basin
2   The cogeneration plant is part of the Würzburg city scene, without dominating it. View of the city center with the Würzburg Residence on the right in the background
3   View from the Old Bridge over the River Main

1

4,5 [4: p. 19] Layers and louvers: aluminum fins in silver and bronze colors form the plant's new skin. The fins' vertical accent is counter-pointed by horizontal cornice strips.

6 One aim of redesigning the plant's surroundings was to make the harbor and the bank of the River Main accessible once more to residents. Terraced seating now steps down gently to the water where coal was unloaded for the power station until 2003.

7,8 Color scheme inside the power station

Good architectural photography is a suggestive art. Wellcomposed photos taken in stark light often remove buildings from the realities of day-to-day perception so much that, upon the first real encounter with these buildings, they often come across like shadows of their own images. If you stand in front of the Würzburg cogeneration plant, you may be surprised to discover that, even on a rainy January day, the building stands there as powerfully as in the photos by Constantin Meyer. Not so much "architecture" as a technological veil zestfully placed around the building, the glinting, silver- and copper-colored power plant forms a new landmark at Würzburg's northern entrance to the city.

At the beginning of the 1950s, the city of Würzburg constructed the power plant directly on the banks of the River Main. Simple, symmetrical industrial architecture was reflected in the water. But over the course of four decades, alterations and extensions were made again and again, until the building had finally lost its form entirely. The city tied a proviso for visual improvement to the planned 2002 expansion needed for the installation of a new turbine. A competition for the design of the facade was initiated. Brückner & Brückner Architekten won the first prize, only to discover, shortly thereafter, that the funds for their design—2.2 million from the total budget of 38 million euro—were actually specified in the budget to cover "architecture and unforeseen contingencies." Luckily, complications failed to develop during the installation of the turbine, leaving the budget untouched for the architects.

But how do you reclad an entire power plant for 2.2 million euro? The architects decided on angled metal shapes, to be coated in custom colors—silver and copper—and furnished with pre-drilled holes by the contracted steel fabricator. The vertical angled shapes each consist of silver- and copper-colored strips of sheet metal screwed together along the tip. All the metal sheets are sized identically; the only difference is the angle in which they are screwed together, so that a wave of angled shapes—some rather flat, others tapered to a point—wraps around the building. Rectangular steel tubes that are fixed to the existing exterior walls on the structural grid serve as a substructure. The aluminum fins establish an image of strong verticality, which is emphasized by the three silvery smokestacks. Horizontal bands that stretch across the entire length of the building form a counterpoint to the vertical grid. Since they correspond to the built volumes behind the envelope, their spacing is varied. They unite old and new: from the core of the power station from 1950 to the most recent construction phase. In this way, the envelope transforms the heterogeneous arrangement of the building into a dynamic collage of moldings and layers.

The selected colors, silver and copper, playfully visualize the energy theme; the copper tone on the south and west sides of the fins, for example, is reminiscent of the wire coils of transformers. Well-chosen colors alone, however, do not begin to explain the particular expressiveness of the building. It works because the entire form resembles a large machine: broadly stretched out and thereby soothing, but at the same time vibrating with energy. In

UNFORESEEN BEAUTY

FACADE DESIGN AND URBAN INTEGRATION OF THE WÜRZBURG COGENERATION PLANT
BRÜCKNER & BRÜCKNER ARCHITEKTEN,
TIRSCHENREUTH/WÜRZBURG

sunshine, the envelope on its north side dissolves completely in a silver glimmer. Abstraction is not only a concept here; it becomes physical reality—an experience perceptible to everyone, full of unexpected visual relationships. Whoever looks at the power plant from the old Main bridge, for example, can observe how the abstraction of the facade corresponds to the geometric patterns of the vineyards in the distance.

The effect of an elegant veil laid over the austere technology reaches perfection through the sweeping cantilever of the envelope toward the harbor basin. This curving "prow" is indebted to the proportions of the whole—and leads to the urbanistic aspect of the project: Until 2003, there were large coal bunkers on the north side of the power plant, from which the coal delivered by barge was transferred into the power plant. The irrevocable changeover to gas-fired operation made

ANGLED FACADE Ground plan and cross section of system

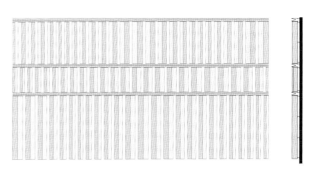

10 m

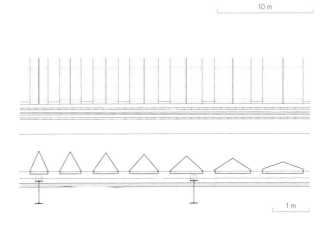

1 m

these coal bins superfluous. Brückner & Brückner, who, a few years earlier, had already converted the neighboring harborside warehouse into the Kulturspeicher museum, recognized the chance to make the harbor, and the section of riverbank that had previously been blocked by the fenced-in power plant, accessible again for Würzburg's residents. And more: The now-functionless harbor basin with its backdrop of disused harbor cranes seemed ideal as an open-air cultural site. Hence the architects transformed the outdoor space beneath the protruding level of the power plant into a spacious terrace that runs down in wide steps to the waterline along the harbor. In the summer, concerts are held here on a floating stage, films are shown, and meanwhile a floating gallery has shown up next door. Where coal was still being unloaded in 2003, summer evenings today come to a close with cold drinks. FPJ

PROJECT DATA
Client Heizkraftwerk Würzburg GmbH
Construction costs EUR 2.4 M net
Enveloped surface area of facade
approx. 10,000 m²
Total quantity of angled shapes 720
Quantity of screws approx. 60,000
Gross volume 113,500 m³
Feature After installing a new turbine, the existing cogeneration plant is clad with an envelope of aluminum sheets—the reconfiguration lastingly improves the surroundings of the power plant.
Completion 9/2006
Project management/team Stephanie Gengler, Stephanie Sauer
Location Veitshöchheimer Straße/ Friedensbrücke, D–97070 Würzburg

Year of construction 1950

1000 ———————————— 2010

Conversion 2007

2000 ———————————— 2010

Construction costs   **Price per m²**
2.4 M €   240 €

15,000
10,000
5,000
0

PIER ARTS CENTRE

— Reiach and Hall
Architects

1   Gable, sheet metal cladding, glass and fieldstone—the architectural elements of the extended Pier Arts Centre are shown here in an architect's sketch.
2   The appearance of the existing buildings is characterized by the traditional fieldstone of the Orkney Islands.
3   The entrance to the Arts Centre is located in a white building on Victoria Street, away from the harbor.
4   The harbor side of the gallery

6

7

9

8

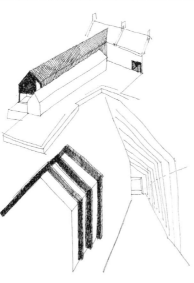

5–7 The new exhibition building
illuminates its surroundings at night
like a lantern. The ribbed structure of
the roof emphasizes the transparent
lightness of the building envelope.
8   Architect's concept sketch
9   Zinc siding at twilight: facade
detail facing the harbor

THREE BUILDINGS FOR ART

RENOVATION AND EXTENSION OF
THE PIER ARTS CENTRE, ORKNEY ISLANDS, SCOTLAND
REIACH AND HALL ARCHITECTS, EDINBURGH

To someone from the south of Scotland, Stromness is located far to the north—in a place that seems more Scandinavian than Scottish. The architects Reiach and Hall view their work in the context of a northern modernism. They aim to create architecture characterized by stillness, lightness, and clarity. Stromness, the second largest city on the Orkney Islands and their main port, is distinguished by a unique seaside urban development—with stone piers, old warehouses, and mercantile buildings, all of which give the rugged, windswept seafront of the town its character. The Pier Arts Centre (PAC) occupies a strategic position in this stone frontage, directly adjacent to the ferry terminal. The Centre houses one of the most significant collections of British art from the twentieth century, and the permanent collection is regularly supplemented by temporary exhibitions.

Established in 1979 in two landmarked historical buildings directly on the stone pier, the Pier Art Centre was refurbished and expanded by Reiach and Hall Architects from 2005 to 2007. The ensemble consists of three distinct elements: a building along Victoria Street on the landward side and two buildings that extend perpendicularly from the street towards the sea. The whitewashed streetside building contains the entrance, offices, and library, along with an artist's studio. The latter two are a renovated waterfront warehouse and a new building in black with a simple pitched roof in a modern form that recalls traditional warehouses. The original pier building contains the permanent collection, whereas the temporary exhibitions are held in its newer counterpart.

Like the renovated existing buildings, the new building designed by Reiach and Hall has become part of the urban topography—and yet its black exterior defamiliarises the seemingly conventional form and elevates it subtly from its historical surroundings. Unlike the waterfront warehouses of stone, this building has an exterior that oscillates between solid and void: black zinc ribs alternate with translucent glass infill. At night, the illuminated interior radiates like a lantern from between the ribs of the outer skin.

Ragna Róbertsdóttir, an Icelandic artist who works with volcanic material, inspired the architects to create a facade toward the sea that appears to shift as the viewer moves. The annex takes the same basic form as the older building alongside it, and the spacing of the ribs in its glass facade echo the original gallery's rafters. When viewing its gable wall head-on, the building appears solid, yet with each step the viewer takes to the left or right, this solidity dissolves. Seen from the side, the new building merges somewhat into the background, allowing the more massive, original, pier building to regain prominence. The new building signifies stability and strength, but in its lightness and transparency, it contrasts refreshingly with the hard stone walls of the neighboring buildings.

PROJECT DATA

Client Pier Arts Centre, Stromness
Construction costs GBP 2.8 M
Usable floor area 658 m²
GFA 1,023 m²
Completion 2007
Architecture/management Neil Gillespie, David Anderson
Location Victoria Street, Stromness, KW16 3AA Orkney, UK

Year of construction 1820

1000 ——————————— 2010

Conversion 2007

2000 ——————————— 2010

Construction costs	Price per m²
approx. 3.4 M €	5,167 €

15,000
10,000
5,000
0

LONGITUDINAL SECTION THROUGH THE ENSEMBLE

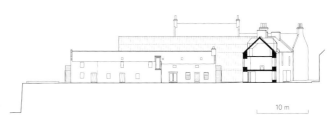

10 m

GROUND FLOOR PLAN

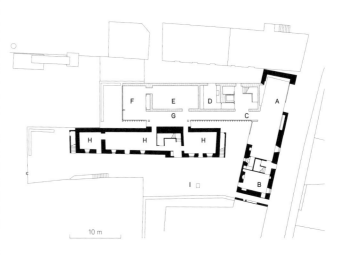

10 m

A Entry/meeting/shop	E Gallery-double height	I Hepworth curved
B Administration	F Gallery	form Trevaligon
C Long gallery	G Pend link	
D Workshop	H Collection	

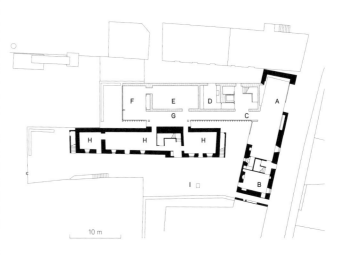

The situation on the pier before the new building was erected

The Victoria Street building can be seen as the antithesis of the black house: here everything is white. The inviting, friendly impression of its white walls makes it seem familiar, yet upon closer inspection, this building also has an uncanny air about it. Gray and brown tones dominate the town of Stromness, and in front of this backdrop, the Pier Arts Centre's vibrant white main building seems no less strange than its black counterpart.

The interior spaces of the gallery are plastered white and punctuated by glazed openings in the gable wall and along the side facing the sea. Spacious, tranquil exhibition spaces enable the works of art to come across at their best; soft northern light filters into the spaces and establishes a connection between the visitor and the surrounding landscape. FPJ

10   The gallery offers a view directly
onto the harbor and the fishing boats
of Stromness.
11,12 Inside the gallery

CAFETERIA IN THE ZEUGHAUS RUIN

KASSEL (DE)
OLD Ruin —pp. 38, 42
NEW Restaurant/Bar —pp. 92, 136, 156, 168

— Kassel Building Department
Hans-Joachim Neukäter

1, 4 The architects have inserted the cafeteria between the 500-year-old walls as a dematerialized glazed volume, whose real walls are those of the former armory.

2 The ruin as it stood in the 1960s

3 In 1972, two-thirds of the Renaissance building were demolished to make way for a new school.

5 All that remains of the Zeughaus ruin is the southwestern corner—seen here in its entirety from Artilleriestraße.

6–9 The cafeteria has been inserted as a free-standing structure within the old armory walls, yet it is closely interlinked with the existing substance. The area inside the ruin is reached across a small terrace.

FROM OBSTACLE TO CONNECTING LINK

A NEW CAFETERIA IN KASSEL'S ZEUGHAUS RUIN
KASSEL BUILDING DEPARTMENT,
PROF. HANS-JOACHIM NEUKÄTER

The matter-of-fact manner in which the city of Kassel went about demolishing the external walls of its war-damaged, historical armory—known as the Zeughaus—is appalling from today's perspective: Photos from 1972 show an excavator ripping into the meter-thick walls—it is hard to believe that, not all that long ago, the demolition of a Renaissance building was still the order of the day.

On March 1, 1582, Landgrave Wilhelm IV laid the four cornerstones of the Zeughaus, in which the city's arsenal of weapons was housed. "The building, whose considerable dimensions are highlighted in older descriptions, shall be on plan a rectangle of 96.80 meters long and 21.80 meters wide," according to a historical source.

At the beginning of the 1970s, the city planned to construct two vocational schools on the Zeughaus site. For one of them, the Max Eyth School, two-thirds of the long, rectangular building was demolished. After the completion of both schools, four-story classroom blocks were attached to the corner section of the ruin on two sides. "They probably would have liked to tear down the entire ruin," comments architect Hans-Joachim Neukäter, in retrospect, about the decision to retain a portion. "Actually, the ruin was even more disruptive at that point. It blocked the connection between the two wings!"

However, awareness of the city's architectural heritage had grown. In 1991, Kassel citizens founded an association with the goal of preserving the ruin and giving it a new use. Association members freed the overgrown inner area of debris and financed the refurbishment of the external walls with donations. In order to achieve permanent use of the building, Hans-Joachim Neukäter, director of the municipal building department, presented a design that received unanimous consent and was ultimately realized: A cafeteria, missing from the school until now, was to be integrated into the ruin. The newly created space is at the disposal of the vocational students and is also available for public events.

The architect describes his concept as follows: "Inside and outside correspond to old and new. The outside is sturdy, the inside is fragile. This gave birth to the idea of integrating a glass object into the ruin, one that does not restrict the space or make it smaller, and does not compete with the imposing appearance of the remaining outer walls. Three horizontal layers characterize the design concept: Level 1 of the cafeteria (ground floor), which lies half a meter above the former floor, the set-back mezzanine level above, and the roof slab. Level 1 and the mezzanine level connect the two classroom wings, accommodate their differences in height, and are enveloped solely by a transparent glass and steel construction." Neukäter consciously chose a mezzanine instead of a second full story so the inserted new building remains as light as possible and permits, from nearly every point, a view of the fieldstone masonry of the external walls—which he perceives as the real shell of the building. The architect folded the thermal envelope at the top of the building to the inside, again with the goal of interrupting the view of the external walls as little as possible. The few direct points of contact between the masonry ruins and the new building have been detailed meticulously. The unbuilt-upon part of the ruin above the cafeteria terrace is now once again accessible.

Structurally, the building's foundation slab rests on grillage supported by bored piles. The historical floor, made of large format, 15 cm thick sandstone slabs, lies 60 cm below the terrazzo floor of the new building—the present ground floor—and remains untouched. Integrated underfloor heating makes additional heating elements superfluous. The number of columns carrying the reinforced concrete slabs for the mezzanine and the roof has been reduced to a minimum. The exterior enclosure is formed by triple glazing supported by a delicately profiled post-and-rail construction.

The cafeteria and the ruin complement each other while maintaining their own identities. That's the way the landmark preservation office also saw it, giving the design concept its full support. FPJ

LONGITUDINAL SECTION THROUGH THE OLD ARMORY AND THE CAFETERIA
with the floors of the vocational school on the right

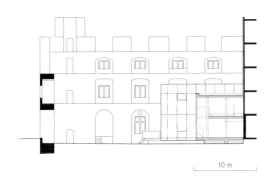

10 m

FLOOR PLAN AND FUNCTIONAL ZONES OF THE CAFETERIA

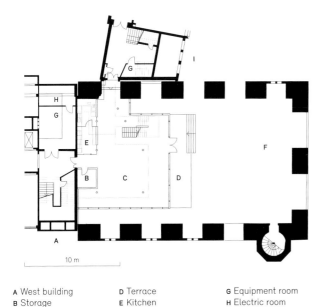

10 m

A West building
B Storage
C Cafeteria
D Terrace
E Kitchen
F Zeughaus ruin
G Equipment room
H Electric room
I South building

In the Second World War, Kassel was the target of aerial attacks by the Allies; following bombing in 1943, the Zeughaus was gutted by fire—only its external walls remained. In the following decades, the shell deteriorated and became overgrown. Alongside plans to reconstruct the building as Kassel's museum of local history, for some time consideration was given to using it as a parking garage. Ultimately, though, none of the plans were pursued further.

PROJECT DATA

Client Kassel municipal administration
Construction costs EUR 1.14 M
Usable floor area 257 m²
GFA 305 m²
Feature The new building is glazed all around; the shell construction corresponds to 80% of the total built volume.
Completion 12/2008
Project management/team Claus Wienecke, Margitta Heidenreich
Location Artilleriestraße/corner of Zeughausstraße, D–34125 Kassel

Year of construction 1582

1000   2010

Conversion 2008

2000   2010

Construction costs   Price per m²
1.14 M €   4,436 €

15,000
10,000
5,000
0

MUSEUM KUNST DER WESTKÜSTE
— Sunder-Plassmann Architekten

1 The main exhibition hall in the reconstructed barn; the light comes through a "slit" cut in the ridge, where natural and artificial light mix.
2 The three exhibition halls are linked by views through them.

2

3

4

3–5 Inside Grethjens Gasthof:
On the basis of its historical remains,
the architects have re-interpreted
the building demolished in the last
century.
6   View from the gallery into the hall

5

6

A NEW CENTER FOR THE VILLAGE

RECONSTRUCTION IN A RURAL CONTEXT
SUNDER-PLASSMANN ARCHITEKTEN, KAPPELN

Strictly speaking, the Museum Kunst der Westküste (West Coast Art Museum) does not really belong in a book about building within the existing fabric, because, apart from two outbuildings, the ensemble is purely new construction. If you put aside the conventional distinction between existing buildings and new construction, this museum on the North Frisian island of Föhr manifests itself as a particularly interesting example of old and new: The old part being referenced here is not a physical building, but the history of the site, located at the heart of a village. The existing fabric, in other words, is first and foremost imaginary, and that is what makes the museum perhaps the most interesting contemporary approach to the subject of architectural reconstruction.

Both of the main buildings, which delimit the ensemble toward the street, are reconstructions of buildings that previously existed on this site: The first exhibition hall, in the form of an indigenous thatch-roofed barn, stands on the site of a barn of this type that existed here until 1968. The entrance to the museum is located in a narrow passage, enclosed by the clay-daubed brick walls of the barn and the radiantly whitewashed inn. As a visitor, you rub your eyes somewhat in disbelief, because even this building, which from a distance is a lovely, old inn with equally old linden trees in front of the door, is a new building. It is a reconstruction of Grethjens Gasthof, an inn that was a meeting place for artists working on the Frisian west coast around 1900.

The museum has been founded by a pharmaceutical entrepreneur, Frederik Paulsen, whose father came from Alkersum. His aim in doing so was to make his extensive collection accessible to the public. Over decades, he collected paintings depicting the sea and the coast, made along the North Sea coast between Holland and Norway. Alongside maritime subjects of prestigious artists like Max Liebermann, Edvard Munch, and Emil Nolde, he focused on the "west coast painters," a group centered on Otto H. Engel, who were active around the turn of the century on Föhr and elsewhere. Their most important meeting place was the aforementioned inn: Grethjens Gasthof.

The client wanted a museum that would fit into the context of a village of four hundred inhabitants, conform to international museum standards, and serve as a rendezvous for visitors and locals alike. This bundle of almost contradictory aspirations was a welcome challenge for Gregor and Brigitte Sunder-Plassmann, from Kappeln in Schleswig-Holstein. The couple, an architect and an art historian, are comfortable working with north German building types: They possess considerable experience of complex building tasks and have built, or renovated, nearly two dozen museums in the past twenty years. The design phase was preceded by a meticulous study of local materials, building volumes, and building typologies. Instead of a solitary, self-contained new building for the museum, the architects developed an ensemble of seven buildings, two of which already existed on the site. In order to make the former den of the "west coast painters" tangible once again, they reconstructed the inn, which had been demolished in the intervening years, on its existing masonry foundations. This serves today as the museum cafe and an event venue. Behind the attractive inn building are the spacious museum garden and six buildings of different sizes that accommodate exhibition space. The largest of these assumes the form and materials used in old Frisian barns. The ensemble as a whole is laid out in such a way as to recreate the village center, which was demolished in the 1970s.

In the exhibition halls, the architects have used a combination of daylight and artificial light. The "barn," the large exhibition hall that bounds the ensemble on its eastern side, is top-lit by daylight entering through a slit cut along the ridge of the roof. Spotlights are also installed along the inside of the ridge opening, so that daylight and artificial light blend into one source of light, as in both of the adjoining halls.

The architecture critic Ulrike Kunkel views the museum as a balancing act between creative reconstruction and modern architectural language. She concludes that "in a rural context, even a partly historicizing architectural language can be the appropriate one." FPJ

CROSS SECTION THROUGH THE EXHIBITION HALLS with the barn on the left

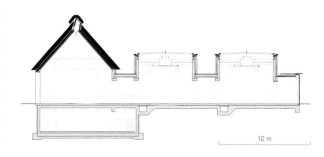

10 m

THE MUSEUM ENSEMBLE AT THE HEART OF THE VILLAGE OF ALKERSUM

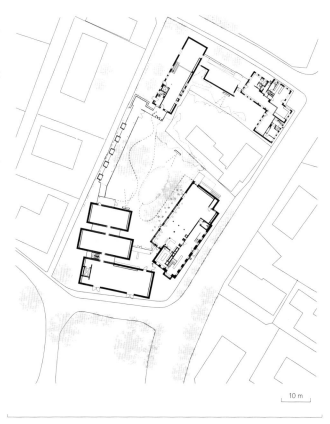

10 m

PROJECT DATA

Client Municipality of Alkersum/Föhr and Nesos GmbH, Frederik Paulsen, Alkersum
Construction costs EUR 13.2 M
Usable floor area approx. 5,600 m²
Gross volume 10,304 m³
GFA 1,950 m²
Completion 7/2009
Project management Gregor Sunder-Plassmann
Construction management Thomas Paulsen, Wyk/Föhr
Location Hauptstraße 1, D–25938 Alkersum/Föhr

Conversion 2008

2000   2010

Construction costs	Price per m²
13.2 M €	2,357 €

15,000
10,000
5,000
0

7   Access to the museum's garden
8,9   Nature in the mirror: glass walls
give the museum's garden a sense of
spaciousness.

10,11 Ticket desk and museum shop,
interior and exterior views
12   Exterior view of the ensemble:
The entrance lies between the
thatched barn and Grethjens Gasthof.

ROOF FOR THE KLEINER SCHLOSSHOF

— Peter Kulka Architektur

1,2 Between Renaissance and Baroque: the modern dome within the roofscape of the rebuilt Dresden Royal Palace, view from the Zwinger side
3 Axonometric view of the dome and its connection to the existing roof
4 [p. 40] The Kleiner Schlosshof with its new roof

1|2

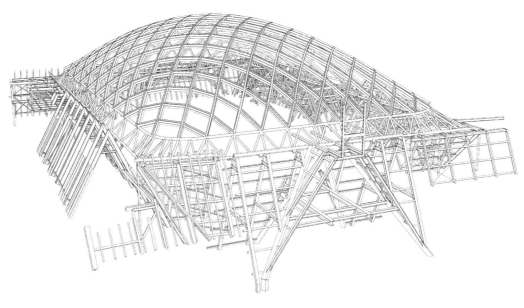

3

AIRY DESIGN

DOMED ROOF FOR THE COURTYARD
OF DRESDEN'S ROYAL PALACE
PETER KULKA ARCHITEKTUR, DRESDEN

The Royal Palace in Dresden has been the residence of the Saxon kings since 1547. In the Allied bombing raid of February 13, 1945 that destroyed Dresden, the palace burned down to its masonry foundations. The exterior reconstruction of Dresden's Royal Palace began in 1986 and was concluded in 2006. The Saxony state government had previously resolved to use the rebuilt palace to house the museums of the Staatliche Kunstsammlungen Dresden (Dresden State Art Collections).

Every day, the Dresden State Art Collections receive several thousand visitors. As the number of visitors grew, questions arose about where the museum's affiliated infrastructure, such as the foyer, ticket desks, and information facilities, could be best accommodated. A study of alternatives showed that the centrally located courtyard, known as the Kleiner Schlosshof, was best suited for this purpose. In roofing it over, a design was sought that would still allow people to experience the courtyard facades—with their pediments, small towers, and varying cornice heights—in their entirety. A difficult undertaking: different variants, especially glass domes, were scrutinized. However, all the options investigated would have either required structural reinforcement of the recently completed reconstruction, or would have considerably impaired the spatial effect of the Renaissance courtyard. To avoid obstructing views of the manifold architecture, a structure was needed that would be supported near the ridgeline of the surrounding roofs.

After years of discussion, a proposal by the architect Peter Kulka was chosen. It envisaged spanning the palace courtyard with a self-supporting, trussed dome. The 84-ton steel domed roof is formed as a load-bearing lattice shell that is based on a network of rigid nodes. The dome is curved in both the transverse and longitudinal directions and makes it possible to cover the 615 sq m courtyard without intermediate supports. It spans a maximum of 45 m and measures 8.35 m in height, from the circumferential truss at its base to the peak. This pillow-shaped form—charming despite its simplicity—has proven to be an attractive contemporary element on the skyline of Dresden's historic center. Its geometry was developed in a three-dimensional computer model and was designed using CAD to enable precise assembly down to the last millimeter. In order to avoid elaborate and costly measures for reinforcing the existing buildings, it was particularly necessary to reduce the horizontal thrust of the dome and ensure the least possible dead load.

Riggers with mountain-climbing experience filled in the rhomboid interstitial spaces of the 1,400 sq m dome with 265 pneumatic pillows. These are made of a transparent, very light, and weather-resistant plastic (ETFE membrane), which has been tested under extreme conditions in the desert of Arizona. The pillows are kept under a constant internal pressure of 800 Pascal. Their supply air comes directly through the cavities of the load-bearing steel sections—hollow square tubes with a cross section of 18×18 cm, making a secondary system for air supply superfluous.

The pneumatic pillows up close

There were several arguments in favor of this unusual roofing solution and against a steel-and-glass structure: The membrane pillows minimize the weight of the structure—directly, due to their low dead load, and indirectly, because the steel structure supporting them requires thinner cross sections. Only in this way was it possible to support the structure via the roofs of the surrounding buildings. The rhombus size is 4.1×2.8 m along the diagonal axes of the roof—enabling the courtyard to receive a maximum of daylight and preserving the effect of the former exterior space. The architects chose the rhombus for their vaulted roof shell because the diamond shape is a motif rooted in Renaissance architecture. If it had been executed in glass, the dual curvature of the dome would have necessitated dividing every "diamond" into two triangles. Complicated junctures and a dubious appearance would have been the consequences.

PROJECT DATA
Client Free State of Saxony
Construction costs EUR 7.5 M
Usable floor area 615 m²
(floor area of Kleiner Schlosshof)
Coverage of dome approx. 1,250 m²
Surface area of membrane roof
approx. 1,400 m²
Completion 1/2009
Project management Peter Kulka,
Philipp Stamborski
Team Christoph Goeke, Egbert Heller,
Thorsten Mildner
Location Schlossstraße,
D–01067 Dresden

Year of construction 1547

1000 | 2010

Conversion 2009

2000 | 2010

Construction costs	Price per m²
7.5 M €	12,195 €

15,000
10,000
5,000
0

SECTION THROUGH THE ROYAL PALACE with the Grosser Schlosshof in the middle and the Kleiner Schlosshof at the right

JUNCTION OF THE DOME with the palace roof

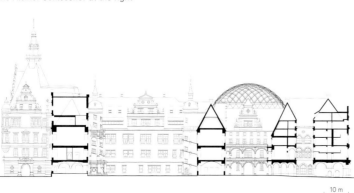

10 m

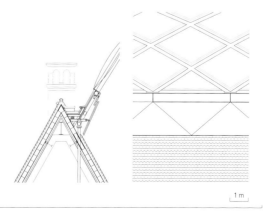

1 m

ENSCHEDE CULTURE CLUSTER
— SeARCH

1–3 An eye-catcher by day and night: the new, six-story tower on the factory site—the glistening metal mesh of its outer skin is finely woven and so light that it billows in the wind.

4   The silhouette of the reconstructed area
5   Six angular, three-story terraced
houses have been newly constructed.
6,7   Views of the collection in
Twentse Welle regional museum

PHOENIX IN CHAIN MAIL

CONVERSION OF A FORMER TEXTILE FACTORY INTO THE "CULTURE CLUSTER"
SEARCH, AMSTERDAM

ANNEKE BOKERN

On May 13, 2000, a fireworks factory went up in flames in Enschede in the Netherlands. The explosion of many tons of gunpowder killed twenty-two people and injured hundreds more. Almost the entire neighborhood of Roombeek was destroyed. Along with numerous residential buildings, industrial landmarks from the old textile city's heyday also fell victim. Only the Rozendaal factory avoided serious damage. This textile factory, built in 1907 in the center of Roombeek, was considered to be of secondary importance and was actually designated for demolition. But since it was one of the few buildings from the industrial heyday to survive the explosion almost entirely undamaged, it became worthy of preservation.

In the meantime, the conversion of the factory into what is locally known as the Cultuurcluster—with museum, art center, studios, apartments, and café—is finished. The Cluster is located in the center of the rebuilt neighborhood of Roombeek. The architects from the Amsterdam office SeARCH have not treated the built substance sensitively: As envisaged in the urban master plan, they have demolished all the buildings on the west side of the wedge-shaped site, except for the perimeter wall, and have replaced them with new buildings that appear almost sculptural. The once-private factory site is now traversed by a footpath.

Along this path stands a new, six-story tower with wraparound strip windows and a facade of metal mesh that swings freely in the wind, alluding through its textile aesthetic to the history of the factory. The tower is a symbol and point of orientation for the Culture Cluster, but notably accommodates the offices and the foyer of the Twentse Welle museum, which now combines three existing collections on textile history, local history, and natural history beneath a single roof. Temporary exhibitions take place on the ground floor of a snakelike new building that connects directly to the foyer and contains studio apartments on its two upper floors. The natural history and local history collection is housed in a 110-meter-long former warehouse with a sawtooth roof. Located on the east side of the site, it can only be reached through a tunnel crossing underneath the footpath. In the warehouse, steel sliding doors and worn brick walls contrast with the new glass showcases and the sound-absorbing, machine-applied plaster between the skylights. There is plenty of space here

for the regional history collection, which ranges from stuffed wolves and re-creations of Stone Age dwellings all the way to steam engines.

A bright red steel bridge between the warehouse and the tower leads visitors to the exit. At the northern end of the site, a depot building has been retained, now housing an art center. This building actually stands on stilts. In order to gain additional space, the architects closed the interstitial spaces with glass walls. The third existing building to have been preserved is the factory owner's villa at the southern end of the complex,

PROJECT DATA

Client Municipality of Enschede, DMO
Construction costs EUR 22 M
Usable floor area 15,000 m²
Completion 3/2008
Project management/team Bjarne Mastenbroek, Uda Visser, Remco Wieringa, Ton Gilissen, Thomas van Schaick, Ad Bogerman, Wesley Lanckriet, Guus Peters, Alan Lam, Alexandra Schmitz, Fabian Wallmüller, Mónica Carriço, Nolly Vos met Frisly Colop, Michael Drobnik, Noëmi Vos, Bert van Diepen
Location Roomweg/Stroinksbleekweg, NL–7523 XG Enschede

Year of construction 1907

1000 ——————————————— 2010

Conversion 2008

2000 ——————————————— 2010

Construction costs   **Price per m²**
22 M €   1,467 €

PLAN OF THE TERRACED HOUSES (FIRST FLOOR)

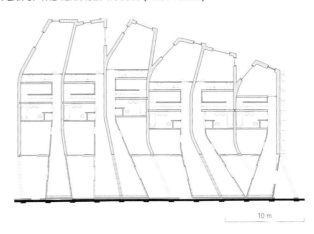

10 m

VOLUMETRIC DIAGRAM OF THE NEW FUNCTIONAL LAYOUT

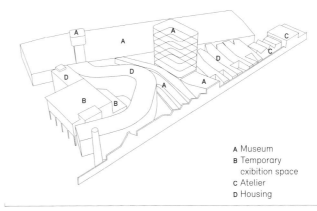

A Museum
B Temporary exibition space
C Atelier
D Housing

After the explosion—the factory site cleared of debris

in which studios are now located. In between, small residential houses stand in a row—unexpectedly—behind the old factory wall. Like the entire complex, this wall resembles a hodgepodge: Over the course of time, window openings have been created or closed repeatedly, additional loading hatches have been inserted, and holes have been filled with new types of brick. This lively, patchwork-like texture, together with its patina, forms a striking contrast to the new residential houses. The houses were spaced so as to fit the factory wall's structural grid. While they appear to be conventional row houses when looking at the wall from off-site, their rears, which face the internal street, are designed as individual, organically formed volumes. This sculptural distortion creates three sharply tapered niches facing the communal street, which have been clad with colorful tiles.

Although the changes to the factory site may have been radical, what emerges in the end is a coherent entity that, with its historical vestiges, brings charisma to the new district.

NEO LEO
VERTICAL LIVING

COLOGNE (DE)
Residential — pp. 22, 32, 42, 50, 62, 84, 132

— Lüderwaldt Verhoff Architekten

VERTICAL DOMESTIC FURNITURE

INSERTED STAIRWAY IN COLOGNE
LÜDERWALDT VERHOFF ARCHITEKTEN, COLOGNE

The clients are owners of a late nineteenth-century *(Gründer-zeit)* apartment building in the Ehrenfeld district of Cologne, and have lived on the second upper floor there for years. The eighty-square-meter apartment was already fairly cramped for a family with two children, so when the tenants on the third floor moved out, the opportunity arose to expand by opening up their living space to the third floor and the attic. An opening in the floor was possible, but the clients were against the idea of a spiral staircase. In discussions with the architects, the idea was developed of retroactively inserting a "proper" stair within the building.

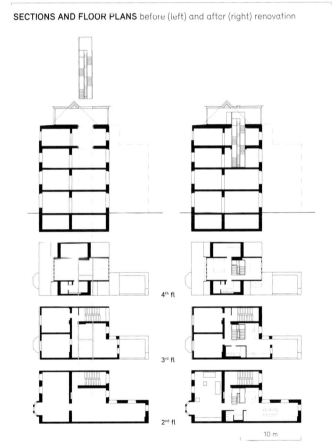

SECTIONS AND FLOOR PLANS before (left) and after (right) renovation

4th fl

3rd fl

2nd fl

10 m

decided upon an unusual, but expedient procedure: The entire stair tower was prefabricated in a workshop and was meant to then be hoisted in one piece through the opened roof with the help of two cranes.

After completing their design, the architects had the CAD drawings revised by a wood construction engineer and sent to the Swiss company that was to cut the panels and other parts to size. The finished components were delivered to a Cologne carpentry firm, where they were completely assembled—proving the fine precision of the preliminary work done by their Swiss colleagues. The surface treatment of the wood, with oil, was also carried out at this stage. The void of the stairwell opening was painted with an orange-red stain, which shows to great effect when illuminated. The steps and landings, which are mortised into the side wood panels, are surfaced with green linoleum. To save weight and provide articulating accents, circular openings were cut in the wood, creating grip holes that simultaneously substitute for the handrails.

In the newly-created shaft, two new steel beams have been inserted at the top of the second floor to provide bearing for the stairway. Structurally, it hangs freely: no loads are transferred to the construction beneath the base of the stair, which begins above the first upper floor. Cleats screwed onto its sides transfer the loads to the steel beams. Hoisting the stair into place was a precision job because the architects allowed for only 1.5 centimeters of tolerance at the sides for the "box." Nevertheless, after an hour and a half the stair was positioned in its designated place and the roof could be closed again.

For the architects, the wooden structure is more than merely a stair—and that's why they have given the stair enclosure a series of additional functions as it passes through the three floor levels—serving as spatial divider, storage closet, parapet, and bookshelf. Moreover, as an orange-red sculpture it creates a design continuum for the three floors and the functional areas located around the new core.

The architect Dirk Lüderwaldt sums it up: "We would like to do something like that again—but unfortunately, another suitable project for architectural furniture of that sort has not ensued." FPJ

PROJECT DATA
Client Private
Construction costs EUR 0.2 M net (total alterations)
Usable floor area 165 m²
GFA 569 m² total rehab, 210 m² alteration
Gross volume 2420 m³ total rehab, 930 m³ alteration
Feature The body of the stairway inserted into the building is formed by 11-m-high Kerto-Q panels fabricated into one unit; Holzbaupreis NRW 2006; Kölner Architekturpreis 2006; Biennale contribution for 2006; Deutscher Holzbaupreis 2007 (shortlisted)
Completion 10/2005
Planing and construction management Dirk Lüderwaldt, Josef Verhoff
Location Leostraße, D–51145 Cologne-Ehrenfeld

Year of construction 1903

1000   2010

Conversion 2005

2000   2010

Construction costs	Price per m²
0.2 M €	1,212 €

15,000

10,000

5,000

0

Accomplishing this within the building's small footprint and without structural complications required minimizing, as much as possible, the weight and need for space. In light of the poor state of the existing floor construction and the diverse floor plan layouts, a freestanding structure proved to be too complicated and thus too expensive. That's how the idea of constructing the stair as oversized built-in furniture came about. The architects studied products and ultimately found the solution in the form of engineered laminated panels (Kerto-Q). The very rugged, stable and large-format panels made of veneered plywood are 56 millimeters thick and up to 11 meters long.

Taking the available space into consideration, the architects designed a stair tower 10.5 meters high with a square footprint of 2.40 by 2.40 meters. The panels, which are connected like furniture using mortise and tenon joints and dowels, form an autonomous, self-bracing structure. Because the outer walls are so thin, the stair itself can occupy nearly the entire enclosed volume. The Kerto panels, however, would be much too cumbersome for assembly within the house—which is why the client and the architects

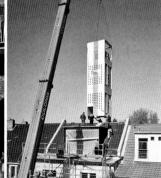

In front of the building, above it, and inside: lifting the staircase into position

2

3  4

1   [p.46] The decisive moment: the preassembled "vertical domestic furniture" is hoisted by crane through the opened roof—and fits at the first go.
2–4   The new element is not merely a staircase, but also shelving, storage space, closet, light fixture, and playground in one.

5   At its base in the second floor;
the staircase has fundamentally
changed the spatial relationship
among the three upper floor levels.

ROSE AM LEND

— Innocad Architekten

1   Flowers emerge from the wall:
the side facade of the "Rose," with
its contemporary interpretation of
stucco decoration

A ROSE IN BLACK

REFURBISHMENT AND EXTENSION OF A BAROQUE HOUSE
INNOCAD ARCHITEKTEN, GRAZ

SIMONE JUNG

Built in the late eighteenth century, with a gable facing the street and the name "Rose" given it by the architects, this is the oldest surviving building in the area. The Rose am Lend is located in the Lend district of Graz, a part of the city which has recently experienced significant change: With its bordellos and bars, it was considered a rather dubious area ten years ago, but the district has since experienced a perceptible improvement, triggered above all by the completion of the extravagant art museum by Peter Cook and Colin Fournier in 2003.

The creative ambience that developed in the wake of the new museum building has turned the district bordering Graz's old town into a popular and slightly hip neighborhood. The residents, though, still represent a colorful mixture when it comes to origin, education, and income; gentrification is only beginning. With their refurbishment and renovation concept for the old building, Innocad Architekten sought to avoid, as much as possible, contributing to price increases and displacement: Acquired by the architects together with partners, the building was refurbished and extended with the help of subsidies from the rehabilitation program of the Austrian state of Styria. Public funding and the distribution of costs amongst the eleven partners of the owner's association has made affordable home ownership possible, as well as moderate rents for units rented to third parties. The social mix is meant to be preserved.

Whereas the late-Baroque old building has been refurbished, the courtyard building from the early twentieth century has been extended by one floor to a total of three stories. This greater density ensures the financial viability of the project, while the altogether better utilization of the site increases the number of dwellings from five to eleven. Along with the interior furnishings, which were individually coordinated with the apartment owners, the landscaping of the inner courtyard contributes to an enhancement of the courtyard situation. The streetside ground floor is still used as a retail shop, but the shoe store that was a long-term tenant has made way for a design and furniture store. The fascinating thing about the renovated house is its facade: The architects have covered it with a black outer skin embedded with glistening granules, making it shimmer in gray or black tones depending on the light. Since the black stucco applied to the exterior

insulation and finishing system (EIFS) could have caused cracks in the facade due to strong heat build-up in the summer, Innocad Architekten, together with the stucco manufacturer, developed a concept that has been used for the first time on this building: a membrane has been added as a separating layer between the black stucco and the underlying thermal insulation, to prevent thermal stress causing cracks in the stucco. Besides its glistening treatment, the homogeneous surface is enlivened by sculptural, projecting rose-shaped applications. Unlike traditional facade ornamentation, the roses, which are staggered and of different sizes, appear to be merely a detail of a continuous, much larger pattern. They give the facade a

PROJECT DATA

Client Golden Nugget Bauträger GmbH
Construction costs EUR 980,000
Constructed area 300 m²
Usable floor area 790 m²
GFA 1,085 m²
Completion 9/2008
Project management Oliver Kupfner
Location Lendplatz 41, A–8020 Graz

Year of construction 1780

1000   2010

Conversion 2009

2000   2010

Construction costs   **Price per m²**
0.98 M €   1,240 €

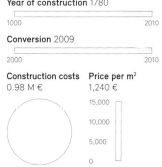

15,000
10,000
5,000
0

The "Rose" and its rear annex before rehabilitation

REAR ELEVATION AND CROSS SECTION

ELEVATION OF THE PAIRED BUILDINGS

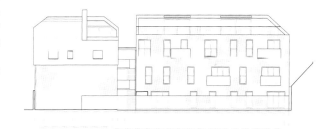

PLAN

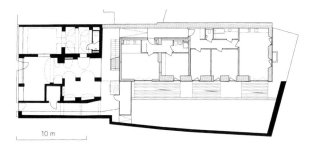

10 m

spatial dimension. In a much smaller form, the roses appear elsewhere—for instance, on the railings of the stairs within the builing and on wrought-iron balcony railings. Beyond these details, the architects have chosen an aesthetic of the greatest possible contrast: the outer skin of the building stands out conspicuously from the surrounding urban landscape.

With the rose motif, as with the name chosen for the building, the architects refer to Saint Rosalia of Palermo, a saint revered initially in Sicily. A golden statue of her stands on Lendplatz, gazing toward the house. Between the historical buildings to its left and right, Rose am Lend, with its black skin, seems surreal and yet familiar: an exotic apparition in traditional form, strange and at the same time offering a moment of quiescence.

2   Courtyard side of the ensemble
3   The street front of the Rose
4   The "joint" between the old
Baroque building and the rear annex
simultaneously provides access.
5,6   The enlarged annex contains
spacious apartments (with open-plan
layouts).

EXPECT SURPRISES
— Three Examples from the Work of RKW Architektur + Städtebau

Building within the existing fabric means planning despite many unknown factors, and requires a willingness to deal with the unforeseen. Since the early 1980s, the architects from RKW Architektur + Städtebau have gathered much experience in the revitalization of existing buildings. Three projects—a palace, a department store, and a school—demonstrate how the processes involved in renovation can be successfully structured.

CARSTEN SAUERBREI

Despite intensive analysis and a thorough investigation of existing conditions, the first surprise comes quickly: It may be that the existing load-bearing structure turns out to be unsound, the waterproofing is contaminated with hazardous substances, or the unheeded basement turns out to have the remains of a medieval vaulted cellar worthy of preservation. All three cases have consequences for the remaining construction process.

"Perpetually updated planning" is what Johannes Ringel of RKW Architektur + Städtebau calls the procedure that leads to success in such cases: Do not stick rigidly to decisions once they have been made, but respond flexibly to new situations instead. If the construction drawings and documentation for an existing building are available, they are especially helpful and can save protracted research. Plans and records from the construction phase do not, however, always provide reliable information—the construction as it was really executed often deviates from the information in the records. This can pertain to the armoring and exact position of utility lines, for instance, or the specified grade of concrete used. A definitive picture does not appear until the information contained in the as-built drawings is complemented with random spot checks of the building. What "perpetually updated planning" means in practice is demonstrated by the three following examples.

SCHLOSS ELLER—PLANNING ACCORDING TO FINDINGS
—

Schloss Eller, a palace situated in southern Düsseldorf, is a testimonial to early Neo-Classicism. It was erected in 1826, presumably by master builder Adolph von Vagedes, in the middle of a landscape park. The master builder integrated the tower of a moated castle from the fourteenth century into the building. The refurbishment concept essentially pursued two goals: On the one hand, the building was meant to be renovated and

CARSTEN SAUERBREI
—

Carsten Sauerbrei, Bachelor of Arts Architecture, was born in 1974 in Berlin. Beginning in 1995, he studied urban planning and architecture in Berlin, Dresden, and Cottbus. From 2000 to 2006, he studied architecture at the University of Applied Sciences in Potsdam (FHP). In September 2007 he began the master's program in architectural education at the BTU Cottbus. Since 2002, he has worked freelance as an architectural guide in Berlin and Potsdam. Since September 2009, he has worked as a freelance architectural journalist and author.

modernized for future use as a venue for private parties, corporate functions, and seminars. On the other hand, the intention was to undo modifications and repair damage that occurred in 1970, when the building was clumsily converted into a fashion school. In this case, the architects were indeed able to refer to the old construction drawings; nevertheless, while demolishing the previously added elements, they repeatedly encountered historical building material that did not appear in the as-built drawings and which made changes in the current design desirable. Three fundamentally different strategies for dealing with the historical building fabric have been pursued at Schloss Eller: documenting and conserving the original built substance—which has, however, subsequently been hidden behind new elements; reconstructing the original state; and lastly, preserving and displaying the condition "as found" and integrating it into the restored building.

PRECISELY IRREGULAR—NEW STAIR IN A MEDIEVAL TOWER
—

The last of these approaches was used most notably for the late medieval tower at the center of the building: The

renovation concept envisaged the tower as a central point of access. For this purpose, a double-flight steel stair with a rectangular plan was to be inserted into the tower. When the old masonry was exposed, however, a multitude of elements worthy of preservation appeared: among them a timber-framed wall from 1823, Gothic window arches, and medieval padstone footings. The plans had to be adapted to this new situation: The architects ultimately dropped their plans for the rigid steel structure and instead developed a reinforced concrete stair with an amorphous, freely arched form that conforms closely to the irregular structure of the existing fabric. Its supports could only be inserted into the wall at locations where the old building material would not be destroyed. Thus the architects had to invert the load transfer of one of the upper stair landings; no suitable direct bearing could be found in the surrounding walls and so this landing was suspended from above instead.

For reasons of cost, however, the historical state was not completely restored in all areas. The original clay plaster from the seventeenth century that was discovered in the upper level of the tower was merely documented by the planners and was subsequently clad with a protective covering of gypsum board.

There was intense deliberation over whether to reconstruct the stucco decoration, which had been damaged during the 1970 renovation to the extent that only fragments of it remained. Refurbishing and supplementing the existing remains would have been possible, but costly. Thus it was decided to make a cast of the preserved fragments using a silicone mold, so as to make an exact copy of the original decor for the new—suspended—ceiling. The new stucco decoration was about two-thirds less expensive than a reconstruction would have been. More detail work and considerably greater effort were needed for dealing with the public spaces facing the terrace—the Prinzensaal and the Prinzessin Luise salon. In the Prinzensaal, some of the stucco decoration was freed of numerous coats of paint that had noticeably reduced

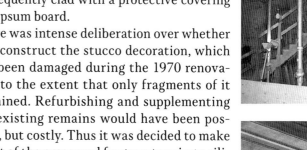

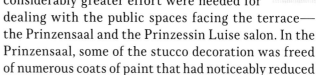

1   [p. 55] Schloss Eller on the southern edge of Düsseldorf

2   The new stair inserted into the medieval tower fits in with irregularities in the existing masonry.

3   The walnut paneling in the Prinzessin Luise salon has been freed of multiple layers of paint and restored to its original condition.

its depth. In the Prinzessin Luise salon, the historical walnut paneling was completely uncovered.

HISTORICAL CELLAR—
MOISTURE ABATEMENT AT ANY PRICE?
—

The initial plan was to waterproof the vaulted cellar, which had been built on a moist subgrade—the part of the cellar located beneath the medieval tower was at one time the ground floor, and in 1823, the grade level was raised by an entire floor with fill. A new, waterproof ground floor slab was discussed as a means to eliminate the moisture, but was ultimately rejected—not least because water could then infiltrate the connection joints with the surrounding masonry walls. The effort and cost would have been too high in proportion to the result, since the planned occasional use of the space for holding formal dinners and wine tastings did not necessitate thoroughly dry walls. Thus the architects decided to use an option that was both simple and inexpensive—they reactivated the old means of cross ventilation. The original window's cross section proved to be sufficient for this purpose. This design decision not only protected the built fabric, it also saved a six-figure sum in building costs. In combination with heating pipes that are installed in the floor and covered with porous precast concrete blocks, the moisture can be carried away through periodic natural ventilation. The salt that is expected to effloresce—as the residue of water vapor diffusion—at the base of the walls can easily be removed in future by routinely brushing it away. A medieval well was found in the ground while carrying out the work, and it is now preserved beneath glass as a reminder of the former moated castle.

STADTPALAIS POTSDAM—
DISCUSS, NEGOTIATE, AGREE
—

For the construction of the department store Stadtpalais Potsdam, the brief was to transform historical relics and building elements masterfully into an entirely new entity. These included the facade and atrium of the predecessor building, an Art Nouveau department

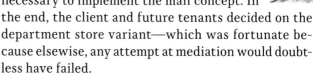

store built in 1907, as well as the remains of an old distillery and a Baroque town house. The complex existing fabric and the fact that the planning could only be completed after starting construction demanded significant negotiating skills and flexibility on the part of the architects.
A fire in February 1996 marked the transition to a second life for the most important

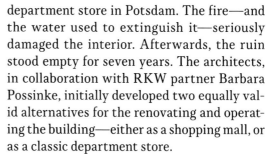

department store in Potsdam. The fire—and the water used to extinguish it—seriously damaged the interior. Afterwards, the ruin stood empty for seven years. The architects, in collaboration with RKW partner Barbara Possinke, initially developed two equally valid alternatives for the renovating and operating the building—either as a shopping mall, or as a classic department store.
Both concepts envisaged preserving as much of the old building fabric as possible. According to Possinke, there was much agreement amongst landmark preservationists to the integration of the retained historical elements. However, they vehemently rejected another measure—moving the historical inner court. This intervention would have been necessary to implement the mall concept. In the end, the client and future tenants decided on the department store variant—which was fortunate because elsewise, any attempt at mediation would doubtless have failed.

REFURBISHMENT OF
THE ART NOUVEAU COURT
—

The extent of consultations and discussions was greatest for the refurbishment of the historical atrium, and especially the reconstruction of the glazed roof. Firstly, the architects convinced the landmark preservation authority that the department store could only be operated profitably if an additional story were built over the courtyard. Then they reconstructed both the color and pattern of the historical glazed roof. Only one quarter of the old glass panes were still available for piecing together the jigsaw, after having survived the fire, albeit blackened from soot and damaged. Originally sporting an Art Nouveau pattern, the glazed roof was renewed around 1920 with a simplified Art Deco design. The architects studied both of these versions together with the landmark preservation agency and they subsequently decided to reconstruct the more recent one. A suitable glass for reconstructing the ceiling coffers was found in cooperation with glass manufacturers and the landmark preservation agency's representatives. The issue of lighting was then addressed because daylight was now obstructed from directly reaching the glass roof by the construction above. In this case, too, a solution closely approximating the original situation was found after an intensive process of consultation.

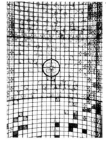

4   [p. 55] The historical facade of the Art Nouveau department store from 1907—now Stadtpalais Potsdam

5   A refurbished old distillery now accommodates a covered market hall.

6   [p. 55] For reconstructing the glass roof in the atrium, the architects only had roughly 25% of the original panes still at their disposal.

7   [p. 55] The ornamentation and colored glass panes were restored in close collaboration with the landmark preservation agency.

CONSTRUCTION PLANNING
DURING CONSTRUCTION

—

"With the project in Potsdam, planning steps normally taken in the first phases of work lasted until the construction documentation was already underway," explains RKW project manager Jan Pieter Fraune. "Ideally, a precise survey of the site being infilled is available at the outset of the planning phase." However in Potsdam, with parts of the building being demolished and others being retained, the digital survey could not be fully conducted until after completing the demolition work. It is indeed possible to conduct trial borings in order to ascertain where the gable or firewall of the neighboring building runs. In practice, though, such conditions cannot be precisely resolved until the building elements being retained have been exposed. Hence the construction planning had to take place in parallel with the construction phase.

FRANZ MEHRING SCHOOL, LEIPZIG—
REGAIN LOST TIME,
KEEP TO BUDGET

—

Things were entirely different with the Franz Mehring School in Leipzig, a prefabricated concrete building from 1973. The pending conversion to a full-day school required multi-purpose rooms and an auditorium, as well as new rooms for music and art lessons. These programmatic spaces were gathered in an annex building, which was clad with green fiber-cement panels.

Preparatory planning work was assisted by the fact that the building department in Leipzig was able to make detailed as-built documents available and possessed experience in the refurbishment of publicly used communist-era buildings. The construction documentation allowed suspected areas of contamination by hazardous substances to be determined at an early stage. The tests showed that such material was only present in the old ventilation ducts: Kamilit, an insulation product made of rock wool—or rather cinder wool—which was common in former East Germany. This was disposed of properly.

CLIENT KNOW-HOW

—

An issue regarding the fire resistance of the ceilings was also quickly resolved with the aid of the client. Nevertheless, conclusions drawn in such matters needed to be continually verified during the construction process, because the plans did not always show what had really been built, according to project manager Jan Fitzner.

This was particularly apparent with the foundation work for the annex. To begin with, the excavation work uncovered underground structural elements and piping that were not plotted on any drawings and these had to be demolished. Consequently, the contractor hired to underpin the existing foundations was not able to finish his work on schedule. Construction delays and extra costs were the result.

Although RKW incorporates extra time for such contingencies into the construction schedule, it was nevertheless a serious challenge to regain time lost by the delay and complete the interior finishings in half a year. To optimize and accelerate the construction work, the architects documented the progress of the work on site every day. They recorded the performance of the contractor with the help of detailed charts and photos. This enabled them to control the deployment of materials and manpower in a carefully targeted manner. Through savings in materials, the budget also ultimately remained within the specified limits.

Creative ways of dealing with restrictions of the existing building required skill. At the outset, the architects wanted to remove entire load-bearing walls from the interior in order to create large, interconnectable spaces. However, as the structural analysis showed, the removal of entire walls would have jeopardized the load-bearing capacity of the building. Hence it was decided to leave small sections of wall to both sides of new openings. The spaces thus created allow much the same scope for use as the original concept.

Considerable skill was required to integrate the new mechanical equipment and modern electrical system that were needed—smoke detectors, electrical outlets, ducts—within interior walls just 10 cm thick. As many lines as possible were shifted to run in the floors and ceilings. Where this could not be accomplished, the architects combined them into the fewest possible ducts. The electrical supply for the corridor lights, for example, goes through a cable raceway that also feeds the classrooms, running

there along the partition wall and below the ceiling to the corridor. The lighting along the side of the corridor underscores the consistent rhythm of columns and beams—a welcome by-product of mounting the light fixtures along the side.

---

8   The final design allowed the Art Nouveau court to be retained, with an additional level being built above it.

9   During the excavation work for the annex, unrecorded underground structures were discovered.

10   Where new supply lines could not be installed under the plaster, they were bundled together in just a few ducts.

11   Spatial sequence after renovation: new openings create a direct connection between the classrooms.

FRANZ MEHRING
SCHOOL
— RKW Architektur + Städtebau

1 Green and compact: the annex provides enough space for everything needed by the school in its new role as a full-day school.
2 Close-up of the fiber-cement panel facade
3 Full-height windows bathe the new building's interior in light.
4 White on white: refurbished rear facade of the school with new solar screening

BALANCE AND COUNTERBALANCE

**REHABILITATION AND EXTENSION
OF FRANZ MEHRING SCHOOL, LEIPZIG**
RKW ARCHITEKTUR + STÄDTEBAU

Franz Mehring Primary School in southern Leipzig was built in 1973 using large prefabricated structural panels to form a slender building typical of its day. The goal of the project, in addition to enabling its future use as a full-day school, was to rehabilitate the existing building. It was necessary to create space for a fourth elementary school class, for the auditorium they previously lacked, and for music and art rooms. The entrance situation was to be improved and the building needed to be made wheelchair-accessible.

In 2005, the architects from RKW Architektur + Städtebau convinced the competition jury not to accommodate the additional spaces and functions in a separate new building, but to house them instead in an extension appended directly to the existing building. This concept was ultimately realized as a four-story annex clad with green fiber-cement panels. Their design is especially convincing in how it deals with the available space: The schoolyard at the rear of the building has been retained in its original size. Circulation routes within the school are minimized because the extension joins at the main corridor of the old building. A new, airy foyer has been added at grade level in front of the existing main entrance. The new spaces can be subdivided and used flexibly.

Along with the extension, the old building was rehabilitated comprehensively: Its exterior walls have been insulated, and the south facade received integrated and flexible sunshading in accordance with the requirements of the EnEV (Energy conservation ordinance) for thermal protection in summer. Colorful panels in front of the columns of the windowed facade add rhythm to the rows of operable windows and emphasize the band-like pattern of the windows.

The idea of positioning the bright green annex directly in front of the austere main facade should not be misunderstood. After all, the new building does not flout the existing one, but develops it further: The motif of horizontal bands that dominates the facade is taken up most notably on the ground floor, which is pulled in front of the broad, gabled front wall as a wall slab. With the school's name inscribed into the stucco, the solid wall characteristic of this type of prefabricated building, or *Plattenbau,* becomes an eye-catcher and thus holds its own against the distinctive annex.

By the same token, there was certainly a desire to establish a counterpoint: A vertically structured building has been added to the functionally designed, rectangular school. With its rounded corners and asymmetrical, slightly wedge-shaped floor plan, the annex appears more organic, standing for individuality, and thereby gives the whole a much more pluralistic expression.

Consequently, the built modification to the school is at the same time an expression of social change: With its classrooms arranged lengthwise and the faculty rooms and ancillary spaces grouped around the main circulation space, the previous building was organized in an eminently functional manner. New educational goals, especially the nationwide trend toward full-day schools, expand the school's mission beyond the mere transfer of knowledge. Thus alongside the after-school care center and the auditorium with its art and music rooms, the musical disciplines are today housed in the extension.

The envelope of the annex consists of fiber cement panels that are pigmented in two custom colors. The inspiration for choosing the two strong green tones was the skin of geckos. Strong colors were also used inside, but the architects chose to refrain from the excesses of those colleagues who give daycare centers and schools a particularly broad and arbitrary spectrum of "cheerful" colors in the name of "child-friendly architecture."

PROJECT DATA

Client City of Leipzig, Building Department
Construction costs EUR 3.8 M
Usable floor area approx. 2,800 m²
GFA 4,980 m²
Gross volume 17,450 m³
Completion 7/2009
Project management/team Norbert Hippler, Jan Fitzner
Location Gletschersteinstraße 9, D–04229 Leipzig

Year of construction 1973

1000	2010

Conversion 2009

2000	2010

Construction costs   Price per m²
3.8 M €   1,357 €

SITE PLAN OF THE EXTENDED SCHOOL

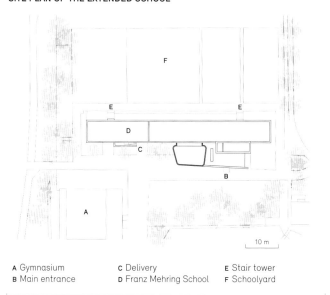

10 m

A Gymnasium   C Delivery   E Stair tower
B Main entrance   D Franz Mehring School   F Schoolyard

MAKING ONE FROM TWO Going from a standardized school building to a functionally extended new facility

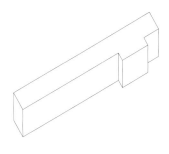
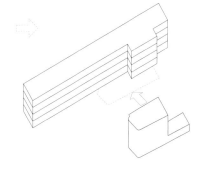
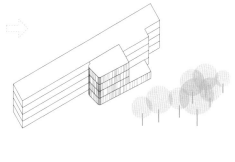

7

5   Direct connections between the
classrooms create good conditions for
instruction in smaller groups.
6   Auditorium and multi-purpose
room, with the stage in the foreground
7   The widened corridor in the
existing building
8   The refurbished main stairway

8

CHESA ALBERTINI

ZUOZ (CH)
OLD Residential
NEW Residential — pp. 22, 32, 42, 46, 50, 84, 132
Gallery — p. 22

— Hans-Jörg Ruch Architektur

FROM HAYLOFT TO ART ROOM

**CONVERSION OF A FARMHOUSE
IN THE ENGADINE VILLAGE OF ZUOZ**
HANS-JÖRG RUCH ARCHITEKTUR, ST. MORITZ

SIMONE JUNG

Engadine houses have special significance within the alpine building tradition. From the sixteenth until the eighteenth century, it was the sole valid building type in the Swiss Engadine region. With the corporeality of its defiant stone walls, it gives the local villages a seemingly urban effect. In addition to its stately style, the close structural and functional coupling of sections for living and working is a key feature: The Engadine farmhouse differs from other rural, single-building farmsteads through the integration of previously exterior spaces like the balcony and courtyard, which also serve as internal streets. These passages—the ground-level *sulèr* leading to the barn and the lower-lying *cuort* leading to the stable—are what make the this house type unique.

In the canton of Grisons today, this building tradition is endangered not only by demolition and decline, but also by real-estate agents: Once the potential of the charmingly aged houses was recognized, they were converted into multi-family residences or hotels with increasing frequency. Large spaces like the *sulèr* or *cuort* are subdivided, traditional rooms like the *stüva* (sitting room), *chambra* (bedroom), or *chadafö* (kitchen) are taken out of their traditional context and converted into individual apartments. In the process, the cultural value of the houses is lost.

One person who has recognized the value of these traditional houses is the Swiss architect Hans-Jörg Ruch, who moved to the Engadine Valley in 1974. Fascinated by the typical buildings of the region, the architect has since converted a dozen of the centuries-old homes of farmers and patricians, driven by enthusiasm for this architectural heritage, but also by great respect for it. For him, working within the existing fabric means making the original essence of a building perceptible. After intensive archeological research, Ruch masterfully places new elements in contrast to the existing building, without infringing on the integrity of the old structures in the process. The new is consequently secondary to the old. New spaces are inserted as visible additions, but modifications to meet modern living habits are kept as invisible as possible.

This is the case with Chesa Albertini, which begins a row of three houses in the center of Zuoz. The core of the building is a double tower structure from the period immediately before the Swabian War (1499). The facade of Chesa Albertini boasts the characteristics typical of rural Engadine houses, such as the asymmetrical arrangement of deeply recessed windows, which vary in size and form within the massive stone facades.

In 2006 Hans-Jörg Ruch converted the Chesa Albertini into a residential house and gallery. The *sulèr* and hayloft on the lower level remain unchanged in terms of their spatial sequence and serve as exhibition spaces. The upper floor is used as living accommodation. The washroom facilities form a clearly discernable, new element that supplements the sequence of historical wooden chambers. They were inserted into the upper story, like houses within the house, beginning in the early seventeenth century. Here, too, Ruch consistently implements his concept of subordinating the new to the old: He decided against converting the chambers themselves into bathrooms, because that would have debased the existing fabric. Instead, the bathrooms were consolidated in a white cube that now stands next to the three wooden bedchambers.

For structural reasons—and in order to clearly separate the dwelling from the gallery—the hayloft walls were extended up to the roof. Where the gallery required a new, geometrically pure structure, this was inserted as an independent element into the existing building, thus making possible its removal at a later date.

The modern materials introduced by Ruch—such as bare iron radiators—fit naturally between the dark wooden planks and crooked stone walls. With the exception of the gallery spaces, where a hardened concrete floor has been selected, all the floors are made of lime mortar or solid larch. The old wood of the chambers and hallways has merely been cleaned with weak lye. No opposition between old and new results, but rather an aesthetic togetherness. An immaterial design element in Chesa Albertini is the light. No light fixtures ought to disturb the spatial effect, so the walls and ceilings are lit with simple, industrial spotlights. In addition to enjoying the art, in Chesa Albertini you can appreciate the unforgettable atmosphere of an Engadine house formed over the course of centuries—an atmosphere that Ruch discerned through all the subsequent layers and then uncovered. He summarizes his standpoint as follows: "I am interested in the spatial experiences within old buildings, and I seek to invigorate special places with my interventions."

PROJECT DATA

Client Monica De Cardenas
Construction costs n.s.
Usable floor area approx. 500 m²
GFA approx. 550 m²
Gross volume approx. 2,500 m³
Completion 2006
Project management Hans-Jörg Ruch, Peter Lacher
Location Via Maistra 41, CH–7524 Zuoz

Year of construction 1499

1000 — 2010

Conversion 2006

2000 — 2010

CROSS SECTION THROUGH THE REFURBISHED HOUSE

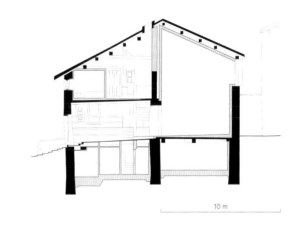

10 m

GROUND FLOOR PLAN

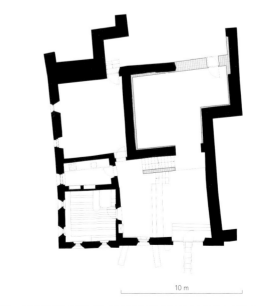

10 m

1 [p. 62] Chesa Albertini in the town center of Zuoz
2 The facade with tapering window recesses
3 The *cuort*: view toward the street
4,5 Exhibition space in the former hayloft
6 Stair to the upper floor

OLD & NEW

7 Upper floor with wooden chambers
8 View from the *sulèr* toward the *stüva* and *chadafö*
9 Stove corner in the *stüva*, framed by centuries-old timber construction

TRANS-
FORMATION

6× ×9,11

12×

×4,10
7× ×5,8 ×3

×1

×2

Transformation [from the Latin preposition *trans*, "across," + *forma*, "form"] denotes a change in appearance, form, or structure: The change embraces a building in its entirety, and the clear boundary between old and new is thereby dissolved. The physical renewal of the built substance is often more subtle, but also more far-reaching, than in expansions. The existing building is newly interpreted in a comprehensive way—frequently enough with entirely different results: In the refurbishment of Stuttgart University's K II high-rise, or the town hall in Weinstadt, the 1960s architectural backdrop remains nearly untouched; the architects have concentrated on emphasizing the strengths of existing aspects and upgrading the buildings to conform with higher technical standards. With the savings bank in Bad Reichenhall and the academic facility in Wildau, in contrast, the architects have recognized that it is worth retaining the physical structure, but not its architectural appearance—and so they have given the buildings complete facelifts.

PUNTA DELLA DOGANA

— Tadao Ando Architect & Associates

ART AT THE TIP OF THE ISLAND

CONVERSION OF A VENETIAN CUSTOMS HALL INTO A MUSEUM
TADAO ANDO ARCHITECT & ASSOCIATES

CLAUDIA HILDNER

The old customs office, Punta della Dogana, occupies a prominent site in the lagoon city: it stands at the tip of an island in the Dorsoduro district, bordered by the Canal Grande and the Canale della Giudecca, diagonally opposite from Piazza San Marco. In the past, investors have repeatedly shown interest in the seventeenth-century building, which has been empty for decades; the changes that would have been necessary for converting it into a hotel or apartment building were, however, not acceptable to the city and its residents. François Pinault, a French billionaire and art collector, was ultimately able to win the Venetians over to his concept for converting the building into a museum of contemporary art.

The architecture of the new "temple of art" is the brainchild of Tadao Ando Architect & Associates. It takes some courage to commission a Japanese architect for converting a European building that is more than three hundred years old: Renovations are not particularly everyday business in Japan, much less the conversion of an existing building that is as old as this. Yet Ando has succeeded in revitalizing the former customs building in a surprisingly sensitive way. It was soon decided to leave the external appearance of the building untouched—not only because of the city authority's strict requirements. The stuccoed brick masonry of the outer walls was carefully restored and, where needed, secured with stainless steel anchors. Slight imperfections in the stucco were repaired, whereas the brick was left visible in larger areas affected by spalling.

The building is covered by a wave-like series of gabled roofs that cover the parallel, long rectangular halls comprising the triangular plan. Atop the restored timber roof structure, the architects have set a new roof that is reminiscent of the original, but which integrates additional skylights.

Inside, to start with, all the partition walls, stairs, and other additions from the last two centuries were removed—only the original structure remains. Walls were left unsurfaced to a great extent. Missing bricks were replaced by the architects only where it was absolutely necessary, using bricks with characteristics coming as close as possible to the existing ones.

The newly inserted structural elements, however, establish a clear contrast to the existing fabric: here Ando has worked with the porcelain-like, polished, exposed concrete that has become his trademark, as well as elements of steel and glass. The floors are—depending on the floor level and the part of the building—either exposed concrete or covered with linoleum. The smooth surfaces contrast with the irregular brick walls and rough wooden beams of the historical building. Since the new and old elements maintain a balance, the existing fabric and the new construction do not compete with one another. Instead, they form a new entity—the art museum. For Ando, this alliance also symbolizes the union of past, present, and future: the shell stands for the past, his architecture represents the present, and the art is that which transcends the present.

PROJECT DATA

Client Palazzo Grassi S.p.A.
Construction costs EUR 20 M
Usable floor area approx. 3,500 m² (net)
GFA 4,331 m²
Completion 5/2009
Project management/team Tadao Ando, Kazuya Okano, Yoshinori Hayashi, Seiichiro Takeuchi
Partner Equilibri srl, Eugenio Tranquilli (project manager)
Location Dorsoduro, Campo della Salute, 2, I–30123 Venice

Year of construction 1500

1000 ——————————————————— 2010

Conversion 2007–2009

2000 ——————————————————— 2010

Construction costs   **Price per m²**
20 M €   5,714 €

15,000
10,000
5,000
0

SITE PLAN AND LAYOUT OF PUNTA DELLA DOGANA

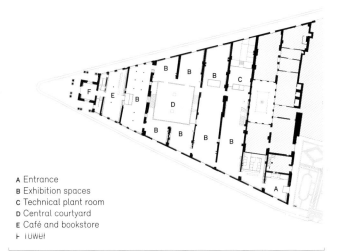

A Entrance
B Exhibition spaces
C Technical plant room
D Central courtyard
E Café and bookstore
F Tower

In the center of Punta della Dogana, the architect has placed a concrete cube that opens upwards. It marks the place where a dividing wall was removed as part of a previous renovation. In this case, rather than reinstating the original structure, Ando has bestowed a new heart upon the building, in the form of this cube. For the flooring of this central exhibition space, the Japanese architect has selected masegni, a large-format sandstone slab that is traditionally used in Venice as paving.

In the end, incidentally, Ando had to dispense with one element of his design: In front of the entrance that faces the Campo della Salute, Ando wanted to position two concrete stelae, so as to call attention to the transformation of the site and its new content. The Venetians opposed it, however, and thus the external sign of the new use remained unrealized.

SECTION THROUGH THE CONVERTED CUSTOMS BUILDING

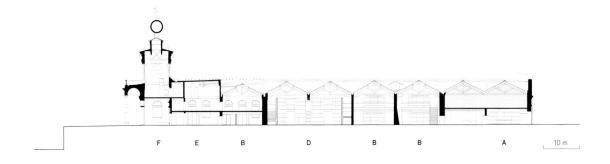

F   E   B   D   B   B   A   10 m

1   [p. 68] The halls in the former
Venetian customs house converted by
Tadao Ando for François Pinault's art
collection are long and high.
2, 4 Square of smooth exposed
concrete—sensitively inserted by Ando
3   Condition before renovation

5   View from the upper level toward
Il Redentore
6   The Canale Grande flows past
the customs house (Dogana da Mar),
which was erected 1678–1682, and
into the open lagoon. The globe carried by Atlas figures once emphasized
the claim to supremacy made by the
Republic of Venice.

ROSSIO RAILWAY STATION
— Broadway Malyan

LISBON (PT)
OLD Train station
NEW Train station
Office — **pp. 76, 116, 132**

1   Modernism from 1887: delicate steel trusswork carries the roof of the main hall.
2   View from the entrance hall toward the tracks
3,5 Escalators connect the raised station concourse with the ground level forecourt.
4   In the course of renovation, 3,000 m² of office space was created. Galleries optimize the use of space.

STATION OF SHORT DISTANCES

REFURBISHMENT AND RENOVATION OF ROSSIO STATION, LISBON
BROADWAY MALYAN , LONDON

From Lisbon's Rossio station, the trains leave for Sintra, a small town not far from the capital, where the magnificent summer residence of the kings of Portugal is found.

The terminal, just a few steps from the lively commercial district of Baixa, is Lisbon's most central train station and also one of the most beautiful in Europe! In 1887, the architect José Luís Monteiro constructed it in Neo-Manueline style for the Portuguese Royal Railway Company. The eye-catchers of the natural stone facade, with its delicate ornamentation, are the two striking, horseshoe-shaped main portals that dominate the plinth facing the square known as the Rossio. The train platform level is almost 30 m above the exits to the street. Escalators and a system of ramps overcome the height difference.

An almost endless row of iron columns carries the station concourse, which is 130 m wide. The elegant supporting structure has been preserved in its original form; at the end of the hall, the train vanishes into a long tunnel.

In 2004, the architectural firm Broadway Malyan, based in London, received the commission to refurbish the landmarked railway station building, after grave damage to the building and in the adjacent railway tunnel had been found. The railway station had last been renovated in 1970. In order to integrate retail shops and movie theaters into the building, several intermediate stories had been inserted without connection to the existing structures or their materiality. Inside, the railway station had mutated into a featureless, labyrinthine building. The goal of the refurbishment, begun in 2004, was to recapture the grandeur of the historical railway station and to restore its once clean and spacious structures. Moreover, the handsome, neighboring Largo do Duque de Cadaval, which until then had been used as an unmanaged parking lot, was redesigned as a car-free urban square. The level of comfort was improved and a welcoming atmosphere was created for travelers, not least by inserting a direct connection between the train platforms and the neighboring Restauradores metro station.

A total of 3,000 sq m of office space and 1,000 sq m of retail and food services have been created inside the railway station, all integrated coherently within the historical context. In the vicinity of the waiting area at platform level (2nd above-ground level), a 1,000 sq m exhibition space has been integrated. The elaborate ornamentation of the light-colored limestone facade has also been meticulously refurbished, as have the surviving cast-iron window frames in the hall.

Dismal and cluttered: Rossio before renovation

The challenge for the architects was to construct an attractive vertical connection, as direct as possible, between the entrances on the Rossio and Praça dos Restauradores on one hand, and the train platforms two levels above, on the other. A wall to the side of the refurbished escalators has been designed with a field of vertical, staggered metal splines. Their wave rhythm enlivens the otherwise sober stairway. In addition to the escalators, a series of stairs and terraced levels connects the concourse with its immediate surroundings. Both elements generate a suitable visual and spatial connection between the city and the railway station proper. The restoration and restructuring of Rossio railway station by Broadway Malyan is convincing in its clarity and in its concentration on a limited spectrum of necessary and effective interventions. After darkness falls, when the newly installed exterior lighting effectively illuminates the Manueline pomp, you realize that the station was built for a king. The excursion to the palaces of Sintra has a worthy beginning. FPJ

PROJECT DATA
Client Invesfer Refer
Construction costs EUR 40 M
Usable floor area approx. 7,600 m²
GFA 8,400 m²
Feature Winner in the "Best Refurbishment" category of the competition held by Portuguese magazine *Construir*, 2008
Completion 2/2008
Project management Broadway Malyan, Lisbon office; Sofia Carrelhas
Location Praça de Dom Pedro IV, P–1100 Lisbon

Year of construction 1887

1000 ▭▭▭ 2010

Conversion 2006–2008

2000 ▭▭▭ 2010

Construction costs   **Price per m²**
40 M €   5,000 €

15,000
10,000
5,000
0

CROSS SECTION THROUGH THE CONCOURSE AND THE SIDE ENTRANCE

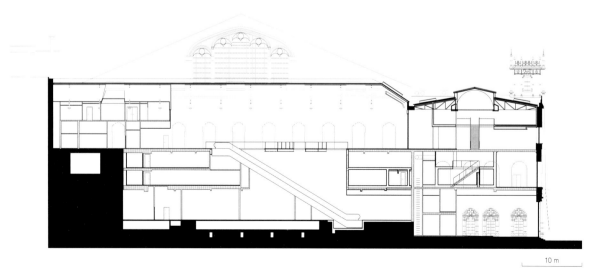

10 m

Light and airy passage:
the entrance hall
7–9 The adjacent public square was
refurbished together with the railway
station—and is now car-free.
10 The Neo-Manueline street
facade with clock tower

SPARKASSE BERCHTES-GADENER LAND
— Bolwin Wulf Architekten

BAD REICHENHALL (DE)
Bank
Office —**pp. 72, 116, 132**

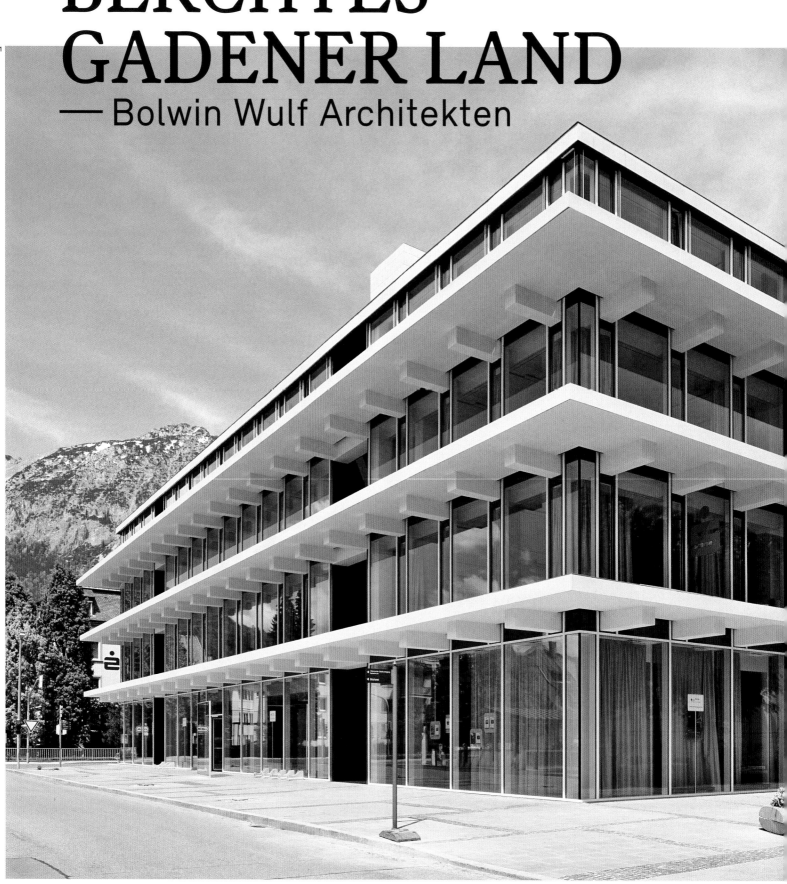

1   Overall view after renovation
2   The internal courtyard, view of
the offices
3,5 Space for bank customers:
entrance, information, and self-
service terminals on the ground floor
4   In place of a stairway that
was no longer needed, a miniature
graduation tower that improves the
building's room climate was created.

6   The refurbished stairway
7,9   Rosé, white, and red: the colors
of salt also dominate the office floors.
8   The renovated top floor serves as
a venue for events.

THE COLORS OF SALT

RENOVATION AND ENERGY-EFFICIENCY UPGRADE, SPARKASSE BERCHTESGADENER LAND IN BAD REICHENHALL
BOLWIN WULF ARCHITEKTEN, BERLIN

The renovated central office of Sparkasse Berchtesgadener Land is the outcome of a competition held by the financial institution in 2006. In addition to structural refurbishment, the competition brief was to restructure the aging main building in such a way that the 235 employees, who had worked until then in several separate buildings, could be brought together beneath one roof.

The architects perceive their design as a strategy of "continued building": the continued use of the building from the 1970s and the "paring away" of its exterior are meant to communicate a consciousness for resources and continuity, particularly since the building possesses evident qualities in terms of its basic structure. Hence the interventions in the building fabric are slight, but of great effect: the original sun shading in front of the facade, which obstructed views both inward and outward, has been removed; the ends of the load-bearing beams have been exposed and horizontal prestressed concrete panels placed on top of them. In combination with the full-height glazing of all the stories, they give the renovated building significant structural clarity. The result is an extremely simple system for sun shading that guarantees maximum transparency and openness. The floor slabs, which cantilever 1.2–1.5 m beyond the facade, protect passers-by from the weather and also correspond to a typical regional element, namely the cornices on the buildings of the historical saltworks in Bad Reichenhall.

The new brief dispensed with the functional mix—bank, commercial, and residential spaces, along with correspondingly complex means of access—that was favored when the building was originally constructed. This made two stairways on the entrance facade superfluous. In their place, the architects have established a trickling water wall along the full height of the building, as a small counterpart to the graduation tower in the neighboring spa park. The small graduation tower of the Sparkasse provides healthy evaporative cooling for the office floors, while serving as a reminder of the spa tradition and the salt production that gave Bad Reichenhall its importance.

Groundwater is used in cooling and heating the building: a geothermal well transports groundwater of a constant temperature to the technical plant. The water provides cooling in the summer, whereas it helps to provide heating in the winter through the use of heat pumps. The spaces are heated or cooled, as needed, through a capillary system integrated into the suspended ceiling; small radiators provide additional individual heat. By installing the heat pump system and improving the energy-efficiency of the building envelope, the net energy needs of the building have been reduced by over 80 percent, thus lowering its CO_2 emissions by approximately 300,000 kg CO_2/year.

The interior of the building has gained enormously from the renovation in terms of clarity, elegance, and freshness. All the spaces are many times brighter than before. Brightness and freshness most notably stem from another of the architects' references to the salt that has been such a formative influence on Bad Reichenhall: the color system of three different pink, violet, and red tones that pervade the building and its furnishings is derived from the natural color spectrum of salt. The salt theme is evident at night, too, thanks to colored accent lighting at each floor level.

Built-in elements that had become superfluous have been removed from all levels. Glazed combi-offices dominate the refurbished office floors. The generous core zones accommodate conference areas, kitchenettes, and all central functions, establishing good conditions for a high degree of both intentional and coincidental communication. Nearly all the built-ins are custom-made.

The new customer lobby and self-service area on the ground floor are bright and open; two recessed inner courtyards allow light into the building's interior. The uppermost level has also undergone reorganization and now has a large, partitionable room for conferences and public events. Many visitors to the savings bank will doubtless fondly remember the spaciousness of its roof terrace and the terrific mountain panorama. FPJ

SECTION THROUGH THE REVITALIZED BUILDING

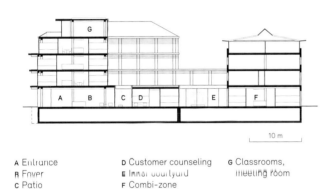

10 m

A Entrance
B Foyer
C Patio
D Customer counseling
E Inner courtyard
F Combi-zone
G Classrooms, meeting room

THE ELEMENTS OF THE THERMAL CONCEPT

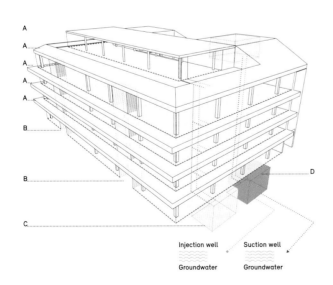

Injection well   Suction well
Groundwater   Groundwater

A Thermally activated ceiling areas for heating/cooling, aided by conventional radiators
B Trickling water wall for natural, adiabatic cooling of internal zones and the staff kitchens
C Gas-fired condensing boiler to accommodate peak loads
D Heat pump for heating/cooling the activated surfaces

PROJECT DATA
Client Sparkasse Berchtesgadener Land
Construction costs EUR 11.3 M (for buildings: EUR 6.2 M; equipment: EUR 2.4 M)
Usable floor area 6,510 m²
GFA 7,230 m² + basement/garage
Gross volume 29,000 m³ + basement/garage
Feature In place of the stairway, the architects have inserted a small graduation tower on the courtyard side of the building, which ensures healthy indoor air in a natural way through evaporative cooling of mineral spring water. Structural sun shading from cantilevered slabs; distinguished with multiple awards, including the Bayerischer Bauherrenpreis Stadterneuerung 2009.
Completion 7/2008
Project management Hanns-Peter Wulf
Location Bahnhofstraße 17, D-83435 Bad Reichenhall

Year of construction 1975

1000   2010

Conversion 2008

2000   2010

Construction costs   **Price per m²**
11.3 M €   3,057 €

15,000

10,000

5,000

0

WEINSTADT TOWN HALL

WEINSTADT (DE)
Town hall

— COAST office architecture

WEINSTADT (DE)
Town hall

DABS OF COLOR FOR THE SIXTIES

RENOVATION AND REFURBISHMENT OF WEINSTADT-BEUTELSBACH TOWN HALL
COAST OFFICE ARCHITECTURE, STUTTGART

Weinstadt town hall was built in 1964. The patinated copper facade and the hexagonal annex with its polygonal roof are characteristic of the light, two-story building. Although it had aged somewhat and was technologically obsolete, there was no doubt about its architectural qualities. Therefore the objective of the refurbishment, in addition to upgrading the energy efficiency and technical services, was first and foremost a spatial reorganization to adapt the building to changes in function and to bring its strengths back to the fore.

The town intended to establish a central citizens' office for all five Weinstadt districts in Weinstadt-Beutelsbach town hall. Additionally, the offices, conference room and wedding room were to be redesigned. The new layout is meant to convey openness and responsiveness to the public.

The redesigned foyer has been articulated with just a few elements, giving visitors clear orientation. Where there was once a dark waiting area, an open foyer flooded with light has appeared. The architects have replaced the cabinet element, which previously separated the offices from the waiting area, with a full-height glass wall. As a result, a visual connection now exists between the entrance and the office area, while the foyer receives daylight from two sides. A light ceiling has replaced the previous dark wooden ceiling, which allows the space to appear more generous. The reception desk has taken the place of the built-in cabinet. Mulberry-colored acoustic panels attract attention to the office zone beyond, and create a pleasant contrast to the otherwise sober ambience of white and gray tones.

A broad stair in the foyer and a new elevator bring visitors to the administrative area on the upper floor. Here too, a white ceiling takes the place of one that was made of dark wood; together with the anthracite-colored floor, it gives the spaces a much ampler feeling. The widened corridor is now more than merely a circulation space. Gallery rails along the top of the walls and recessed downlights in the ceiling have been installed to allow its future use as an exhibition venue for contemporary art.

The assembly hall's characteristic feature is its hexagonal floor plan, with a pointed polygonal roof. This extraordinary spatial effect was the starting point for the architectural concept of reinforcing the impact of the existing elements: The new, white acoustic ceiling now makes the space, which was previously clad with wooden panels, appear considerably higher. Recessed, horizontal lines of light run across the corners of the ceiling, dominating the revamped conference room. They underscore the generosity of the space and ensure uniform illumination.

The demolition of the partition walls in the former offices of the financial department has opened up the opportunity to create a large, continuous space, which can be used as an impressive, ceremonial room for civil weddings, meetings and events.

PROJECT DATA

Client Municipality of Weinstadt, represented by the building department
Construction costs EUR 2.5 M
Primary usable area 1,300 m²
GFA 1,890 m²
Gross volume 6,000 m³
Completion 2006 (first phase)/2009 (second phase)
Project management Alexander Wendlik
Location Marktplatz 1, D-71384 Weinstadt-Beutelsbach

Year of construction 1964

1000 ———————————— 2010

Conversion 2009

2000 ———————————— 2010

Construction costs   **Price per m²**
2.5 M €   1,923 €

15,000

10,000

5,000

0

GROUND FLOOR PLAN

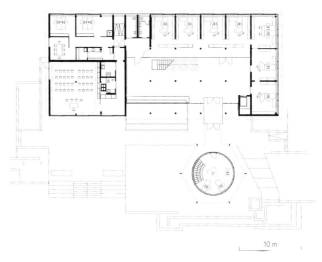

10 m

FIRST FLOOR PLAN

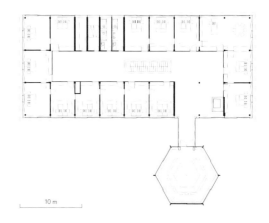

10 m

The new wedding room is reached by an exterior stair from the newly designed forecourt. The ancillary functions—storage rooms, toilets, and the cloak room—are grouped in a room-high, white fixture, whose contour is accentuated by recessed lines of light. The wedding room is defined by ivory and champagne tones—echoes of a white wedding—as well as by the fine texture of the wraparound curtains that frame the actual room. With their different lengths and degrees of transparency, these curtains establish an equally intimate and ceremonial context for a civil wedding ceremony.

The multi-purpose area, however, can also be easily rearranged for everyday use: when the white curtains are drawn back, a spacious, open room for events of every kind emerges. Whatever the occasion, the curtains make it possible to achieve the desired degree of privacy.

The foyer of the town hall before refurbishment

1   [p.80] The council meets beneath a canopy: assembly hall after
renovation
2   Overall view of the revitalized
town hall
3   Workplaces in the citizens' information office
4,5   The foyer of the town hall after
renovation
6,7   White wedding in Weinstadt: the
new wedding room is also available
for other festive occasions.

HEURIED RESIDENTIAL COMPLEX

ZURICH (CH)
Residential — **pp. 22, 32, 42, 46, 50, 62, 84, 132**

— Adrian Streich Architekten

ZURICH (CH)
Residential — **pp. 22, 32, 42, 46, 50, 62, 84, 132**

1   The exterior after refurbishment:
the austere charm of the building's
staggered form remains unaltered.
2,3   The full-height silhouette portraits of construction workers were
replaced with pixelated photographs
of playing children.

4

5

6

4,6   The orange-colored curtains of
the enlarged balconies are popular
with the tenants and fit well with the
curved balconies.
5   The south side of the now-
meandering building

Heuried residential complex was designed and built from 1972 to 1975 by architects Peter Leemann and Claude Paillard. The complex consists of two seven-story buildings that are aligned in a staggered manner typical of their day. They enclose an outdoor space that opens to the south. In other ways as well, the existing buildings were shaped by the spirit of the 1970s: the organic landscaping by Ernst Cramer flows around the angular, large-scale forms, which step back at right angles.

In 2002, the architect Adrian Streich received a commission from the city of Zurich to conduct a study on refurbishing the residential complex. Despite its somewhat austere public image, Streich recognized the formal qualities of the ensemble—its silhouette, the meandering juxtaposition of volumes, and their differentiated coloration. The artistic elements by Edy Brunner and Karl Schneider in the outdoor spaces were no less typical of their day: children's play

THE SEVENTIES IN NEW GARB

REFURBISHMENT AND ALTERATION OF A RESIDENTIAL COMPLEX IN ZURICH
ADRIAN STREICH ARCHITEKTEN, ZURICH

objects influenced by Op Art—a colorful steamship, lollipop-like concrete pillars in the courtyard, a fountain tiled in bright colors—as well as portraits of construction workers covering the height of the facades. These elements were meant to bestow moments of identification to the large-scale architecture.

The architects' goal was to avoid forsaking these features—particularly their close alliance of architecture, art, and outdoor design—in the process of refurbishing the complex.

The energy-efficiency upgrade improved the U-value for the building envelope, which was originally 1.1 W/m2K and is now 0.25 to 0.20 W/m2K. The workers' portraits from the 1970s, however, had to be abandoned in the course of rehabilitating the buildings' facades. At the same time, the architects decided to give the weighty building mass a vibrant and elegant new appearance by integrating new balconies into the facade. This provided an opportunity to add something "soft" to the angular, large-scale, meandering building. Turning its long, staggered front into a succession of waves was seen by the designers as a marvelous opportunity to manipulate the complex sculpturally at the requisite scale. The continuous, cast-in-place concrete slab of the existing balconies was retained by the architects for structural reasons (negative bending moment of the continuous slab), but the old balcony parapets and one third of the existing slab were demolished. The formwork needed for extending the slab was precisely aligned to match the joint pattern of the existing slab. The new, curved parapet elements were placed in the formwork and then cast so as to be monolithic with the existing slab. The enlarged balconies are integrated seamlessly into the new facade and create the impression of a wavelike movement running along the entire development.

As a contemporary equivalent to the imposing facial profiles that originally graced the facade, new, large-scale figurative images have been put on the side facing Talwiesenstrasse—pictures, resolved into dots, of three children playing. With their appearance of having been "blown up," the images appear to change style from figurative to concrete art as the observer comes closer. The murals are based on photographs of artist Judith Elmiger's children. The scanned photos were imported into a CAD program to delineate the individual dots. The vector-based drawing was then sent to a cutting plotter, which produced 1.1 m high and 6 m long strips of adhesive film. These stencils were affixed, piece by piece, onto the stucco, the round "pixel holes" were sprayed with red paint, and the plastic film was then removed.

Judith Elmiger's graphic murals and the wavelike reshaping of the facades through the enlargement of the balconies help to open this previously introverted housing estate to the surrounding neighborhood. Two large projecting roofs extend to the street, underscoring this transformation.

Analogous to the design from the 1970s, the new color concept relies on bold colors: dark umber in contrast with bluish white, luminous orange, and light Bohemian green. The earthy, natural colors of the mineral-based pigments are inspired by the hues of a mountain landscape. They translate the highly articulated buildings into an artificial mountain range and simultaneously establish connections to the neighborhood, where similar colors predominate.

SITE PLAN OF THE RESIDENTIAL ESTATE

50 m

PLAN OF AN ENLARGED BALCONY

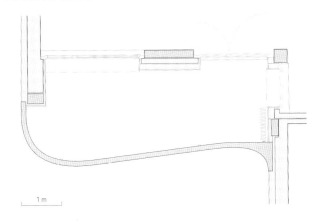

1 m

VIEW FROM HÖFLIWEG

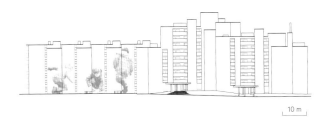

10 m

PROJECT DATA
Client City of Zurich
Construction costs approx. CHF 29 M (building), approx. CHF 34 M (exterior works)
GFA 35,930 m²
Usable floor area 12,563 m²
Feature Energy concept certified in accordance with MINERGIE standard
Completion 2006
Landscape Architecture Manoa Landschaftsarchitekten GmbH, Meilen
Location Höfliweg 2–22, CH–8055 Zurich

Year of construction 1972

1000   2010

Construction 2004–2006

2000   2010

Construction costs   Price per m²
approx. 21.8 M €   1,736 €

15,000

10,000

5,000

0

LESEZEICHEN SALBKE

MAGDEBURG (DE)
OLD Empty lot —p. 32
NEW Urban square

— KARO* Architekten

1   Green living: seating, bookshelves, and a carpet of grass—the
Lesezeichen in Salbke on the out-
skirts of Magdeburg invites people
to dwell a while.

2   The open tower serves as a roof for the small open-air stage and de-limits the site toward the intersection.

3   Before starting construction, KARO* architects simulated the planned building for the residents with a mock-up made from empty beer crates.

4,5   The Lesezeichen stands out prominently from its long-neglected urban surroundings.

6   Salbke's teenagers were allowed to spray graffiti along the base to mark the structure's completion.

A MONUMENT TO PUBLIC SPIRIT

AN OPEN-AIR LIBRARY IN MAGDEBURG
KARO* ARCHITEKTEN, LEIPZIG

Lesezeichen Salbke is a surprising success story: The project, which is in itself modest, became a darling of the architectural press straight after being completed—owing to its potential as an ideal and because it proves that architecture can indeed be a means to effectively redefine places and provide identity. The residents of Salbke have enthusiastically accepted Lesezeichen (meaning "bookmark"), which they helped to design. The square-like ensemble is located at a fork in the neighborhood's main arterial road, on the site of the public library, which burned down in 1987.

With the abrupt deindustrialization following German reunification, Salbke, a one-time village now incorporated into Magdeburg, experienced a drastic decline at the beginning of the 1990s,

SCHEMATIC VIEW OF THE HORTEN TILES
AND THEIR SUPPORTING STRUCTURE

characterized by vacancies and dilapidation. The architect Stefan Rettich describes the initial situation: "As we began the job, we found a neighborhood center that was almost completely abandoned." The municipal administration had barely invested in Salbke since reunification, so perhaps it was pangs of bad conscience that caused them to commission KARO* architects together with Architektur+Netzwerk.

The office, well-versed in urban reconstruction, was given the task of developing concepts for upgrading the image of numerous derelict sites. Given a minimal budget, the task was to demonstrate the potential of these places with temporary, but expressive, actions. The first project was called Wasserzeichen, which means "watermark." On a vacant lot not far from the bank of the River Elbe, sand was laid, colorful flags were raised, and stylized Strandkörbe (canopied beach chairs) were set up. Wasserzeichen was meant to remind people that Salbke lies on the shore of the Elbe, separated from it only by an abandoned industrial site. The idea did not ultimately grow into an urban beach, yet the interest of the residents was aroused.

The opportunity arose to apply this concept of reappropriating urban space to another location, the site of the former library in the center of the neighborhood. In a design workshop open to the public, set up in an empty retail store just opposite, residents young

and old had a week to formulate and sketch out their ideas. In the process, the idea of a "green living room," a combination of urban square and open-air library, came into being, framed by a wall that shields the square from the noise of the adjacent main road. This wall terminates in a two-story, tower-like cube. Toward the street, it visibly identifies the project from a distance; to the square, Lesezeichen opens up in the form of an upside-down "L" that shields a small stage.

To give the participating residents an impression of the spatial effect of their idea, the architects obtained roughly one thousand empty beer crates from a local beverage distributor and used them to build the project as a full-scale, on-site mock-up. The wall of crates was held together with clamps of metal and plastic. For a few days, the residents' group and the architects organized celebrations, readings, and concerts here. "This action brought everyone together," recalled Rettich.

But the funds to build Lesezeichen were nowhere to be seen—until the architects came across the German federal government's Program for Experimental Housing and Urban Development. Their application was successful and the project became a reality. The exterior of the "bookmark" is composed of 50 × 50 cm cast aluminum elements, which the Horten department store group used to clad its buildings from the early 1960s onwards. The architectural firm RKW Architektur + Städtebau had developed the original Horten tiles based on a stylized "H." After the Horten department store in Hamm was demolished in 2006, three hundred square meters of the tiles dismantled went to the Lesezeichen project. The residents' group acquired them, including their subframe, from the demolition company for 5,000 euros. In this respect, Lesezeichen is not an example for building in, but rather with existing fabric. Only the damaged paint finish of the metallic tiles had to be removed and applied anew before they were reused. The facade modules are supported on their original subframe, which is fixed at the top and bottom to a steel frame using steel straps.

Lesezeichen is an idiosyncratic hybrid of interior and exterior space. Its qualities have been achieved with surprisingly economical means, for example the podium-like elevation in relation to its surroundings. Except for the stage and the bookshelves recessed into the wall, the design program is confined to wood-planked benches and geometrical grassed areas. The emblematic presence of the stage tower at the corner of the ensemble was desired by the residents, as was the use of recycled materials. The pure construction costs for the bookmark, including the external works and furnishings, totaled 325,000 euro.

Book donations for the planned open-air library started coming in from the entire city of Magdeburg during the test phase of the 1:1 model, and since then, the residents' group has rented a shop near the site in order to store what meanwhile amounts to more than 20,000 books. A small, non-circulating collection is freely accessible in the shelves of Lesezeichen. Salbke has thus not only won back a part of its old center, but also—unexpectedly—its library.

PROJECT DATA

Client City of Magdeburg
Construction costs EUR 325,000; Budget for planning, residents' participation process, and project documentation: EUR 75,000
Usable floor area 328 m² outdoor area/488 m² site area
GFA 160 m² (stage + urban shelving)
Feature The ensemble is the outcome of a residents' participation process and pilot project for the research program Familien- und Altengerechte Stadtquartiere (Urban neighborhoods for families and the elderly); use of approx. 550 facade tiles from a demolished Horten department store
Completion 6/2009
Project management/team Stefan Rettich, Antje Heuer, and others
Location Alt Salbke 37, D–39122 Magdeburg

Conversion 2008–2009

2000   2010

Construction costs   Price per m²
0.325 M €   398 €

                      15,000

                      10,000

                      5,000

                      0

BACHMANN CONFISERIE
BASEL (CH)
Restaurant/Bar —pp. 28, 136, 156, 168

—— HHF Architects

BASEL (CH)
Restaurant/Bar —pp. 28, 136, 156, 168

PETITS FOURS CLAD IN CHROME

RENOVATION OF A CONFECTIONER'S SHOP
HHF ARCHITECTS, BASEL

HUBERTUS ADAM

After the Rheintor (Rhine Gate) was demolished in 1839–40, the northern historical district of Basel underwent a process of transformation: at the end of the nineteenth century, Marktgasse was built to create a connection between Marktplatz and Schifflände. The street is lined by impressive commercial buildings—such as the building at Marktgasse 4/Blumenrain 1, designed by the Eduard Pfrunder, a busy architect in Basel at the time. The Renaissance Revival building originally housed a restaurant on the ground floor, yet Confiserie Bachmann has been located there for a long time.

With its three shops, Bachmann, family-owned for many generations, is considered an institution in Basel—and the Blumenrain confectioners is both their headquarters and where they produce their sweets. In the 1940s, Hermann Baur established the café and shop on the ground floor, and it was renovated in the 1970s. Until recently, the seating area was dark and isolated from the urban surroundings. Yet Confiserie Bachmann is located on a prominent site: diagonally opposite from Hotel Trois Rois, and only a few steps away from the Mittlere Brücke (Middle Bridge) over the Rhine.

This prompted the family owners to commission a thorough renovation. More light, brightness, and openness—these were the stipulations made to both participants of the limited competition, Jürg Berrel and HHF Architects. In the end, the younger team got the commission.

The renovation by HHF has created an atmosphere that is not typical for a confectioners. They have consciously rejected a plush ambience and sought instead to create a modern, urban venue directly connected to the space outside. The window panes have been extended down nearly to the sidewalk and they've been given borders with a screen-printed pattern of dots to soften the contrast between light and dark somewhat. The windows facing the passageway can be opened outwards in the summer, providing guests who are seated outdoors with some shelter from the wind.

In the center of the space, the undulating display counter of chrome-plated steel is a real eye-catcher; here—as in the display cases near the shop window—chocolates, petits fours, and cakes are presented behind glass, like jewels. Dovetailed with this counter are two more gleaming bars with seating. The rear wall, constructed of two printed glass panes with offset mirror strips, is decisive in establishing the lighting atmosphere in the space. Seen obliquely, the view extends almost indefinitely thanks to the mirror strips, making the interior appear larger than it really is. Seen head-on, the glass construction becomes a luminous but opaque wall with an almost textile character. In addition, there are luminaires from the Italian manufacturer Danese with crimped wiring—some cylindrical in shape and others cuboid with rounded corners. The latter, designed by Carlotta de Bevilacqua in 2007, have been outfitted by HHF with white polycarbonate film and glimmer magically.

The old floor has been retained and solely polished. It gives the light space a feeling of being grounded, in the truest sense of the word, and ultimately makes a reference—as do the dark tables designed by HHF—to the original décor by Hermann Baur.

FLOOR PLAN OF THE RENOVATED CONFECTIONER'S SHOP

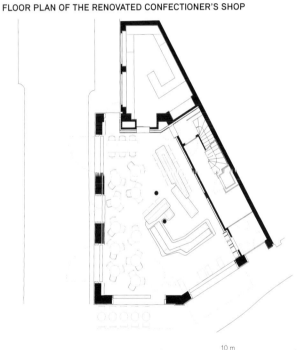

10 m

PROJECT DATA
Client Confiserie Bachmann AG, Basel
Construction costs CHF 1.2 M/EUR 840,000 M net (total alterations)
Usable floor area 139.58 m² (total), café: approx. 98.2 m²
GFA for ground floor 172.82 m²
Feature Special two-layered, back-lit wall surface of sandblasted and printed glass and mirrors
Completion 2009
Planning and construction management Tilo Herlach, Simon Hartmann, Simon Frommenwiler
Project management Markus Leixner
Location Marktgasse 4/Blumenrain 1, CH–4051 Basel

Year of construction 1940

1000   2010

Conversion 2009

2000   2010

Construction costs	Price per m²
0.84 M €	6,000 €
	15,000
	10,000
	5,000
	0

Before renovation, the seating area had a dignified, . . .

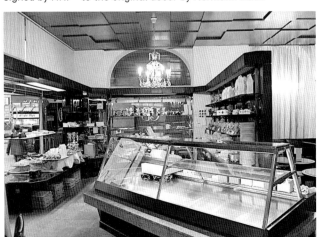
. . . but also somewhat dark and antiquated appearance

2

3

4

1   [p. 92] HHF Architects thoroughly
cleared out the old café. The new
look is provocatively modern, while
conveying clarity and tranquility.
2–4 By opening it to the street and
using light furnishings defined by
few elements, the confectionery is
transformed into a modern café-bar.

SCHOOL IN DAGMERSELLEN
DAGMERSELLEN (CH)
Educational —**pp. 58, 98, 108, 152**
— Peter Affentranger Architekt

2

3

4

1–3   The staircase, designed in red and blue colors, is the new centerpiece of the school.
4   The effect of the colorful reflections is clearly visible.
5   The school's facade and entrance after refurbishment

5

COLORS SUPERIMPOSED IN THE LIGHT

ALTERATION AND EXPANSION OF THE SCHOOL OF HOME ECONOMICS IN DAGMERSELLEN
PETER AFFENTRANGER ARCHITEKT, LUCERNE

After many years in use, the building that houses the home economics school, constructed in 1970 in the Swiss municipality of Dagmersellen, no longer fulfilled the needs of a contemporary school in many respects. It had been in need of renovation for a long time and, moreover, the school urgently needed further classrooms.

Because of the site's exceptional location within the town center, the municipal council invited nine architectural offices to undertake a study. The Lucerne architect Peter Affentranger won the competition.

The new School of Home Economics is intended to mediate amongst the various periods represented on the school campus—a total of six buildings originating from the 1950s and 1960s as well as from the late nineteenth century. The annex, finely tuned in terms of its scale, is well integrated into the ensemble—comprising church, school buildings, and town hall—as a convincing complement. The older parts, about 50% of the current building volume, merge entirely into the new building, which is clearly more generously dimensioned. The existing volume has been reshaped and enhanced. By extending the existing fabric by a fifth wing and an additional story, a harmonious, coherent large-scale form has been created. The stucco-covered facades, with their large windows recessed precisely into the building's shell, enter into a dialogue with the predominantly stuccoed neighboring buildings.

While maintaining the defining structural and spatial elements, the rooms are grouped around the central circulation hall, whose chief focal point is the stairway. At the ground floor, the entrance hall and adjoining library constitute a generous spatial continuum that can be used for events such as readings, concerts, theater performances, and celebrations.

The homerooms and shared classrooms—together with the group activity rooms—form clusters that constitute manageable units on the various floors. To enable forms of instruction and work beyond the traditional classroom, every classroom has been equipped with a window to the access corridor.

The polychromatic color compositions of the interior spaces represent, as it were, a visual intensification of the spatial order—entirely in keeping with the terms of the design, which seeks to create a new whole from what already exists and what is superimposed. In the light, these color tones resonate with one another. The result is a set of well-tempered color spaces that are continually transforming over the course of the day and establishing themselves anew. The harmonious effect of the color compositions, developed by the architects together with artist Erich Häfliger, is not based on the similarity of the colors, but on their characteristic diversity.

The strategy of addition and extension is, in the case of the School of Home Economics, economically sensible and sustainable. Aside from the building shell, which has been newly insulated and executed in accordance with the MINERGIE standard, the functional relationships within the building have been optimized and their diversity of use significantly increased.

Inside and outside of the school building before renovation

PROJECT DATA

Client Municipality of Dagmersellen
Construction costs approx. CHF 6.5 M
Existing usable floor area 1,530 m²
Usable floor area after alteration 2,550 m²
GFA after alteration 2,980 m²
Completion 1/2008
Project management Peter Affentranger, Architekt BSA SWB, Lucerne
Location Kirchfeld, CH–6252 Dagmersellen

Year of construction 1970

1000 ———————————————— 2010

Conversion 2006

2000 ———————————————— 2010

Construction costs   **Price per m²**
approx. 4./5 M €   3,105 €

SITE PLAN AND LAYOUT

A Recess area
B Vestibule
C Entrance Hall
D Handicrafts room
E Library
F Library office
G WC
H Jan.cl.
I Faculty room
J Elevator
K Elevator
L Classroom

10 m

SECTION THROUGH THE SCHOOL BUILDING WITH ADDITIONAL FLOOR

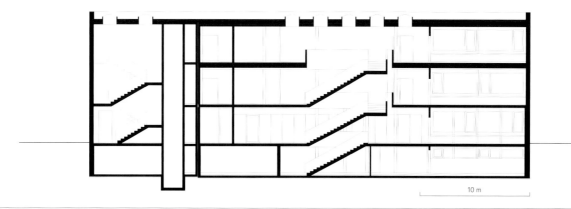

10 m

BLUMEN PRIMARY SCHOOL AND BERNHARD ROSE SCHOOL

BERLIN (DE)
Educational —pp. 58, 95, 108, 152

— huber staudt architects bda

1   The thermal upgrade left the facade at the entrance of the building and its molded concrete blocks unchanged.
2   Aluminum profiles of different widths in detail
3,4   Varying brown tones and profile widths characterize the horizontal strips of the facade, intended to give the impression of wooden slats.
5   Overall view of the refurbished building

SERIAL IN ESSENCE, INDIVIDUAL IN APPEARANCE

ENERGY-EFFICIENCY UPGRADE AND FACADE DESIGN FOR BLUMEN PRIMARY SCHOOL AND BERNHARD ROSE SCHOOL IN BERLIN
HUBER STAUDT ARCHITECTS BDA, BERLIN

FRANK VETTEL

Standardized buildings that were erected in the 1960s and later, which are today in great need of refurbishment, can be found in the inner-city areas of both the former East and former West Germany—and also in Berlin. After decades of neglect, they are being refurbished structurally and in terms of design for the first time since the 1990s.

The starting point in every case is similar: Relatively simple, re-inforced concrete buildings with few special details, whose "egal-itarian" creative indigence has meanwhile been traversed by generations of students, are at last undergoing specific architectural analysis and reconception. Rather than prolonging the existence of the serial product, this experiment concentrates on bestowing a new identity and on treating the users as individuals.

The Blumen Primary School is a prototype built in 1965/66 for school buildings of the series "SK Berlin," which were subsequently constructed hundredfold in East Germany. For this site, the collective goal—both for the architects and for ourselves as clients—was to formulate an individual response to the extant situation. The desire to articulate the facades, so as to give them a hitherto lacking sense of scale, had to be balanced against the usual constraints of public construction, such as the need to work with scanty funds and to reduce long-term operating costs. This led to the concept of an elementary curtain-wall screen that is perceptibly added to the exterior; this becomes a filter and, at the same time, an intermediary between exterior and interior space. After intensive dialogue between the architects and the client, the initial ideas of using an amorphous textile fabric or wooden lattice were revised, resulting in a structure of aluminum profiles in varying sizes and anodized colors from the brown-beige spectrum. Wood was rejected as a building material by the borough, owing to the anticipated maintenance costs.

Mass-produced in a thoroughly serial manner (prefabricated and installed as room-high elements), yet individual in its appearance, the structure provides weather protection for the reinforced concrete facade, which is now properly insulated [cf. p. 102]. The aluminum cladding also marks the separation between classrooms, while functioning as a visual boundary and sun shading in the hallways.

The distinctive horizontal accent of the new curtain-wall facade underscores the massing and the austere poetry of postwar Modernism, and it enters into a dialogue with the trees in front of the building. Aluminum, a material otherwise considered rather cool, demonstrates a high degree of liveliness here through the varied widths of the profiles and their bold, warm colors, as well as through the delicate overall effect of rails, gaps and shadow joints together with the almost painterly refracted light that filters inside. The result is no merely "colorful," run-of-the-mill, elementary school, but rather aesthetically enhanced, urban architecture.

PROJECT DATA

Client Bezirksamt Friedrichshain-Kreuzberg, Berlin
Construction costs EUR 2 M
Usable floor area 1,281 m²
GFA 4,512 m²
Renovated facade 4,900 m²
Feature In 1965/66, the building was the prototype of a standardized school type (SK Berlin) erected in great numbers in the subsequent years.
Completion 2007
Project management/team Andreas Büttner, Stefania Dziura, Leander Moons
Location Andreasstraße 50–52/ Singerstraße 87, D–10243 Berlin

Year of construction 1965

1000 ———————————————— 2010

Conversion 2006–2007

2000 ———————————————— 2010

Construction costs | **Price per m²**
2 M € | 1,561 €

15,000
10,000
5,000
0

ELEVATION AND SECTION OF THE FACADE CLADDING

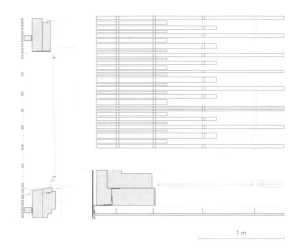

1 m

GROUND FLOOR PLAN

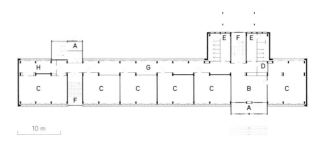

10 m

A Vestibule   D Superintendent   G Corridor
B Foyer   E WC   H Prep room
C Classroom   F Stair tower

View of the school before renovation

When the school was originally being constructed, the artist Norbert Schubert was commissioned to create a facade sculpture for an end wall. It shows the Soviet cosmonaut Yuri Gagarin—an idol in the mid-1960s for all children in East Germany—in the guise of Icarus. In the course of the energy-efficiency upgrade, the architects were able to find the artist and win him over to reconstruct his work on the newly insulated facade; another bit of continuity with the Modernist era.

"Great expectations were placed in the sustainability of the facade refurbishment. The contrasting colours help to prevent graffiti, and the new appearance has led to a re-evaluation of the schools. Internally, too, the small-scale detailing has found broad acceptance." Joachim Staudt [*Detail*, issue 9/2009, p. 898]

TOP MARKS FOR ENERGY CONSERVATION
— Refurbishing School and University Buildings

Up-to-date insulation, efficient building technology, and above all intelligent refurbishment concepts, are helping to transform even forty-year-old prefabricated-panel buildings into low-energy buildings, without changing their external appearance. Here, we discuss experience gained from the modernization of three post-war school and university buildings.

SUSANNE REXROTH

In the years between 1950 and 1970, Germany experienced a construction boom as part of the economic growth of the post-war years. Many of the buildings constructed during these years of economic recovery later turned into "problem children"—irrespective of their formal qualities—and there were two main reasons for this: During the post-war period, building materials were distinctly lacking in quality and hence durability; in addition, industrial building products (in western Europe) and industrial construction methods (in eastern Europe) began to dominate in the early 1960s. Especially during the early years, this industrialization of the construction process had not really been developed to full perfection.[1] For this reason, a large proportion of the buildings from this phase of increased construction activity has now reached the end of its service life and no longer complies with the repeatedly changed energy standards.

Likewise, requirements for comfort, interior fitting out, and infrastructure are no longer the same as those of forty or fifty years ago. New organizational and communication structures, changes in user needs, and requirements for building services make it necessary to modernize these buildings and remove certain building materials, such as those containing asbestos, which have since been classed as hazardous to health.

In school buildings, the high number of pupils in the classrooms leads to very substantial heat loads, which means that air quality standards are much harder to achieve. In most cases, the air quality in existing school buildings is unsatisfactory. In addition, in view of the need for pupils to increase their learning and concentration abilities, natural and artificial lighting are today considered to be much more important.

Also, a number of the school buildings that were constructed prior to 1977, i.e. before the first Thermal Insulation Ordinance came into force, only have inadequate thermal insulation in their building envelope, if any at all. This is where an upgrade of the existing building can sometimes achieve greatest energy efficiency.

MODULAR SCHOOLS IN EAST GERMANY: BERNHARD ROSE SCHOOL AND BLUMEN PRIMARY SCHOOL IN BERLIN-FRIEDRICHSHAIN

—

Owing to the scarcity of materials and low production capacity, East Germany was forced to adopt more economic construction methods soon after the war. Uniform room schedules were developed for schools and apartments, which made it possible to adopt modular types of construction for buildings.[II] In 1965/66, Gerhard Hölke developed a structural frame system for the Berlin State Housing Department (Wohnungsbaukombinat Berlin). Known as SK-Berlin, this introduced a new construction method: a type of modular school, of which more than 160 were built throughout East Germany. The prototype was Bernhard Rose School in Berlin-Friedrichshain, which as an example of this mass-produced modular construction is now classified by the borough as worthy of preservation.

A primary school, Blumen School, was built close by, using the same type of modular construction. In 2008, Huber Staudt Architekten from Berlin were commissioned with the energy upgrade of the facades of both schools. The borough of Friedrichshain was able to fund the work with the help of subsidies. Subsidy programs such as those provided by the KfW Bank for Public Subsidies[III] stipulate strict energy conservation standards when funding projects. In the case of the two schools in Berlin-Friedrichshain, this meant that the values specified by the Energy Conservation Ordinance (EnEV) had to be bettered by 40%, with the result that the annual demand for primary energy has dropped from 27.6 kWh/m³a to 11.4 kWh/m³a.

The upgrade included continuous insulation of the heat-transmitting areas of the building envelope, which had a positive effect on energy conservation, particularly in the main facades of the schools. The end walls of the long, reinforced-concrete block are without windows. These were upgraded at moderate cost, using a conventional sandwich insulation system, which made it possible to improve the U-value to 0.22 W/m²K. The energy upgrade of the stairs and the other facades proved to be more complex. Because the stair facades are constructed with specially shaped concrete blocks to create a decorative effect, the architects opted for internal insulation, consisting of 8 cm foam glass. This improved the U-value from 1.01 W/m²K to 0.4 W/m²K.

The sides of the school building consist of a ribbon facade with spandrel panels made of fair-faced concrete rib

elements with core insulation; this makes up the largest part of the building envelope. Its energy upgrade therefore had the greatest effect on the overall energy balance. In order to protect the thermal insulation against moisture damage from condensate that might form in the cavities of the construction, the architects installed drainage pipes. Back ventilated cladding was installed on the facade elements, consisting of 12 cm mineral fiber insulation, a moisture-diffusing membrane, and aluminum sections. The result of this upgrade is a U-value of 0.20 W/m²K compared to the old U-value of 0.69 W/m²K. The roof parapet was similarly insulated, resulting in a U-value of 0.23 W/m²K, the same as at the plinth, which was insulated with 12 cm perimeter insulation. These measures contribute significantly to the retention of the original character of the facade behind the new "curtain." The aluminum sections were fitted to the entire west-facing courtyard facade of Blumen Primary School, allowing additional solar screening to be dispensed with.

An energy upgrade of the windows was also essential for a successful refurbishment. In order to be able to retain the existing appearance of the staircase windows, which are single-glazed with thin steel profiles and are fitted directly into the concrete wall panels, internal window sashes were installed behind them, producing the effect of double casement windows. This simple measure alone was enough to halve the originally very poor U-value of about 5 W/m²K. The old windows in the classrooms with a U-value of 3.5 W/m²K were replaced with new timber windows with a U-value of 1.3 W/m²K. This meant that the U-value achieved was even better than that required by the subsidy provider, by a margin of twenty four percent.

The two objectives of reducing the consumption of energy and of improving conditions in the classroom were achieved at a very economic investment of 75 Euro/m³ of building volume.

ROLANDSTRASSE PRIMARY SCHOOL IN DÜSSELDORF

—

The primary school in Rolandstraße in Düsseldorf can be considered an equivalent West German example of the same era. Increasing birth rates throughout the postwar years led to an acute shortage of classroom accommodation in the early 1960s, so the development of new school buildings became an urgent need. In the context of a research study commissioned by Düsseldorf Building Department,[IV] the architect Paul Schneider-Esleben

---

1   [p.104] Firewall of Blumen Primary School after refurbishment: The Icarus motif was also repaired as part of the envelope upgrade.

2   [p.104] Stairway after refurbishment: The interstice in the new linked-casement window also serves as a display case in the context of art teaching.

3   The newly applied insulation layer with fixings for the facade cladding.

built Rolandstraße Primary School in a two-wing, double-H format in 1961. The self-supporting facade of this concrete-frame building features different designs to suit respective room functions, is located in the same plane as the supporting structure and follows a grid pattern that is determined by the uprights and decks.

The general refurbishment of the building, which was carried out from 2004 to 2006 whilst teaching continued, focused primarily on thermal insulation and the modernization of the building services. Since the school building is protected as a landmark, the commissioned architects, Legner und van Ooyen, had to retain the existing appearance and avoid—as far as possible—interfering with the building fabric.

Nevertheless, they succeeded in upgrading the building in accordance with the Energy Conservation Ordinance (EnEV 2002) in force at the time. At an average investment of 1,128 Euro/m² of usable area, the cost of the refurbishment was relatively modest, bearing in mind that this included the expensive removal of harmful substances. Because the building contained materials and components that are hazardous to health, such as asbestos, synthetic mineral fibers (SMF), and polychlorinated biphenyls (PCB), it had to be stripped back to its structural frame.[V] Prior to refurbishment, measurements to determine the presence of PCB revealed levels of more than 300 ng/m³ in some areas. These values were clearly in excess of current limits.[VI]

Following the removal of the harmful substances, the architects applied insulation to the structural members, which were originally located on the outside. Today, prefabricated glass-fiber-reinforced concrete elements encase all potential thermal bridges, such as external supports, the faces of concrete decks, roof parapets, reveals, and lintels. The glass-fiber concrete elements are held together with agraffe-type brackets, which are not visible from the outside, so that the original appearance is retained. The new back-ventilated cladding provided sufficient space for 80 mm of mineral fiber insulation, which reduced the U-value of the originally un-insulated facade from 1.8 W/m²K to 0.45 W/m²K. The interior columns were lined with excelsior board and plastered. The window elements were then installed flush with the finished facade. On the sides of the building, the exposed-aggregate concrete panels in front of the sanitary rooms on the ground floor were replaced with 2.5 cm prefabricated glass-fiber concrete elements with a similar appearance and feel. On the two floors above, the white panels underneath the windows were replaced with vacuum insulation panels. As a result, the U-value

of the opaque part of the window elements was improved to 0.45 W/m²K.

The windows in the classrooms were replaced by double-glazed windows with commercially available frames. The ventilation openings in the opening sashes and fanlights were sized so that the maximum air flow speed does not exceed 0.08 m/s. Ventilation of the rooms now relies on manual window operation, as the non-adjustable cross-ventilation slots in the corridor fanlights that were previously installed are no longer permitted under fire safety regulations. The new louvers in front of the fanlight panels of the windows were fitted for landmark preservation reasons—their only function is an aesthetic one. The ventilation slots in the partition walls to the corridors, which had been installed in order to allow cross ventilation between classrooms, were removed and filled in for fire safety and acoustic insulation reasons.

The external solar screening, a subsequent addition to the original building, was removed as part of the refurbishment. Legner und van Ooyen's design provides for electrically operated Venetian blinds in the gap between the glass panes, allowing the original flush facade to be re-established and thereby complying with another important landmark preservation requirement.

For structural reasons, the external loadbearing prefabricated concrete posts and beams of the fully glazed stairwells were completely removed and replaced by a new steel-and-glass facade. Whereas previously the glass panes were simply fitted between the structural elements and a fiber cement panel on the inside, today's insulating glazing consists of toughened safety glass that is thermally separated from the external steel profile. Likewise, the intermediate floor slab supports have been fitted with insulation and sealed against moisture ingress. Although this slightly changes the appearance of the facade, it nevertheless retains a streamlined simplicity with its few elements. The concealed fixing of the glass panes does not have the benefit of building code approval and for this reason needed special exemption.

The gap between the blue-glazed facing bricks and the concrete fire wall at the classroom wing ends was insulated with blown-in water-repellent mineral insulation. This improved the U-value from 1.3 W/m²K to 0.43 W/m²K. The narrow strip left between the facing brickwork and the corner of the building was surfaced with 20 mm insulating render.

---

4   Rolandstraße Primary School in Düsseldorf: Supporting structure for the thermally optimized facade elements

5   [p.105] The school yard shortly after completion of the school in 1961

6   [p.105] The school designed by Paul Schneider-Esleben, after refurbishment: view of school yard

7   [p.105] The building is characterized by rectangular facade panels which are bordered by narrow shadow joints.

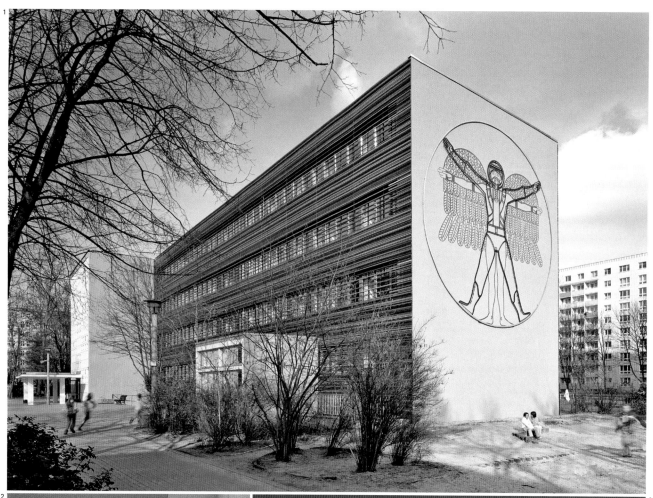

Similarly, the thermal performance of the originally uninsulated roof was upgraded by adding a bitumen-faced sloping insulation layer with a total thickness of between 280 and 390 mm. This reduced the U-value of the roof from 1.5 W/m^2K to 0.26 W/m^2K. The downspouts that had previously conducted rainwater from the roofs of the classrooms on to the stair roof were replaced with a separate, internal roof drainage system, which significantly reduced the risk of water damage to the flat roof.

Both the walls and decks of the cellar had suffered from heavy moisture ingress due to inadequate damp-proofing and lack of thermal insulation. To address this problem, the walls were stripped, water-proofed, and insulated, and the decks were lined with a non-combustible thermal insulation material. The heating requirement of the school dropped to 505 kWh following the refurbishment, which made it possible to re-

SUSANNE REXROTH
—

Dr.-Ing. Susanne Rexroth, born in Karlsruhe, is an architect in Berlin. She studied architecture at the TU Berlin. After working in several architectural offices, she was a research assistant at the Faculty of Architecture at Berlin University of the Arts (UdK), then spent four years in research and teaching at the Institute of Building Construction of the TU Dresden. Since 2009, she has taught and conducted research in the course on Environmental Engineering/ Regenerative Energies. Her main focus is on energy-efficient building and building-integrated solar technology. Her doctorate is on the subject of building-integrated photovoltaics; she has been involved in numerous publications and research projects.

www.f1.htw-berlin.de/studiengang/ut/

underground auditoria were ready for hand-over.

In order to accommodate a varied range of functions within the limited volume available, the architects used a clever device: the story height on the north side is different to that on the south side of the building. By arranging three floors of offices and institute rooms with lower ceilings opposite two floors with lecture halls and studios with higher ceilings, they managed to maximize the available space in the building. As a result, the window sills and concrete spandrel bands on the south elevation are mainly arranged horizontally, whereas the facades on the north, with the lower institute rooms, feature a pattern of individual panels. The narrower east and west facades were constructed in fair-faced concrete with the exception of the corridor areas in the upper stories, where groups of three windows al-

move one of the five gas boilers. Improvements were introduced to heating controls, distribution pipes, and radiators, as well as the lighting system, in order to conserve energy and improve the operation of these systems. The new standard fit-out consists of energy-saving three-band fluorescent strip lights and light fittings with electronic ballast. Together with movement and presence sensors and a daylight-sensitive control mechanism they optimize the use of daylight and minimize the need for artificial lighting.

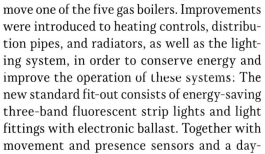

ternate with groups of three concrete panels.

The complex with the two teaching blocks is one of the most typical examples of the architecture of the new Stuttgart School, which was particularly influential between 1946 and 1970. The architects of K II, Gutbier, Wilhelm and Siegel, attempted to achieve a synthesis of the building's formal design with its inherent structure. For this reason they avoided any kind of unnecessary formal embellishments and relied on just three basic materials to define the form of the building: Exposed concrete with different surface finishes for structural elements and facades, fair-faced brickwork for non-loadbearing walls, and timber for interior elements.

BLOCK K II OF STUTTGART UNIVERSITY
—

Stuttgart University, which is located in the inner city, lost many of its buildings during World War II. In spite of the rapid reconstruction of destroyed and damaged university buildings, the growing intake of students could only be accommodated by the construction of Block I, a tower block designed by architects Gutbier, Wilhelm and Siegel, and completed in 1960. Even while the building was going up, it became clear that it would not provide nearly enough space. For this reason, work started on a second, similar block: K II.

After four years of construction work on K II, beginning in 1960, the concrete framed building and its

In the refurbishment of the K II block, which was carried out between 2007 and 2009, the architects Heinle, Wischer und Partner aimed at retaining this design principle while complying with the requirements of the Energy Conservation Ordinance 2002 (EnEV 2002). The architects were able to concentrate primarily on the refurbishment of the interior and the building services, as the facades were only replaced on the north side of the building. There, the aluminum windows were fitted with thermally broken profiles and double glazing, which reduced the U-value to 1.3 W/m^2K. All other facade components—aluminum windows, prefabricated concrete spandrel panels,

8   [p. 105] **Street elevation of the school building in Rolandstraße,** photograph from the time of construction

9   [p. 104] **Night-time view of the K II block of Stuttgart University.** In the background is the K I block, which was constructed a little earlier.

and exposed concrete units—were either repaired or just cleaned. On the east and west elevations, it was possible to retain the existing sandwich construction (apart from the joint-sealing compound, which had to be renewed), which consists of exposed concrete on the outer face with 40 mm cork slabs as core insulation. The integrated heating elements were decommissioned, however, and now additional opening sashes improve night-time cooling. The south-facing facade was fitted with a new, electrically-operated, solar screening system as well as a manually-operated anti-glare system. Once the internal lining of the spandrel panels had been removed, internal insulation consisting of 40 mm foam glass was integrated into the panels. Since the building fabric of the south-facing facade complied with current construction code requirements, the only work required was to clean the existing aluminum windows using an abrasion process and to carry out necessary repairs.

This building, like the majority of its contemporaries, contained substances such as asbestos, PCB, and poly-nuclear aromatic hydrocarbons (PAH), which are now classified as hazardous to health. All components containing asbestos, such as ventilation ducts, fire doors, fire dampers, and machine-applied plaster were removed, as were joint-sealing compounds containing PCB and materials containing PAH in the parquet, roof structure, and the existing cork insulation.

Refurbishment of the building services systems was also extensive: The electrical and data wiring had to be completely re-installed, including the emergency lighting and fire alarm systems. Today the lighting consists exclusively of fluorescent strip lights and compact fluorescent lights. An electrical installation bus (EIB) system controls lighting in the foyer, corridors, and staircases in accordance with the time of day and daylight availability, in conjunction with movement sensors to allow for lighting on demand.

The auditoria were equipped with a ventilation system with heat recovery. The old direct-current motors of the lift system were replaced by gearless synchronous motors, so the former transformer room became

redundant and could be used for the ventilation system. The other rooms are ventilated and cooled using a manually operated night-time ventilation system, which significantly reduces the energy requirement for ventilation and cooling.

All heating/cooling ducts and pipes were insulated in accordance with EnEV requirements; in addition, controlled pumps had to be installed. On the upper floors, all ceiling-mounted heating systems were removed and replaced with radiators underneath the windows, which helped to improve comfort levels significantly.

Following the refurbishment work, the transmission heat loss H'_T of 3.5 W/m^{2}K dropped to 2.5 W/m^{2}K, the energy input was reduced from 1,470 MWh to 1,200 MWh, and the primary energy requirement was reduced from 1,500 MWh/a to 805 MWh/a (forecast for 2010).[VII] These values demonstrate that the energy upgrade has resulted in improvements in the energy balance that are commensurate with the investment volume. This was extremely modest, amounting to 654 Euro/m^2 of usable area for building work, 250 Euro/m^2 of usable area for heating, ventilation and sanitary installation work, and 218 Euro/m^2 of usable area for electrical installation work. There is no doubt that additional investment in the energy upgrade would have resulted in further improvements in the energy balance. However, for various reasons—not least landmark preservation considerations—the designers did not make use of all technically possible improvement options.

These three examples show that it is possible today to modernize and upgrade buildings from the 1960s and 1970s to a standard that nearly meets current building requirements. In this way it is possible not only to reduce energy consumption and running costs, but also to improve user comfort and the quality of workplaces. Preserving such examples of post-war modernist architecture warrants the effort; after completion of the refurbishment work, the remarkable quality of the interiors and the unique pattern language of this long-underestimated epoch can once again be appreciated. In almost all cases, the investment is also commercially viable; usually refurbishment is significantly cheaper than demolition and new construction.

I   Examples are the large-panel, prefabricated buildings of the former East Germany: Although they had been designed for a heating energy consumption ranging from 120 to 160 kWh/m^{2}a, depending on the type of panel used, the consumption in buildings without any upgrade tends to be much higher in practice, up to 220 kWh/m^{2}a. The reasons for this are physical and technical defects, which are usually due to faulty workmanship.

II   Buildings made of prefabricated components were not only constructed in East Germany, but also in western European countries. The external walls were lined with facade cladding or, later on, consisted of prefabricated construction systems such as two- to three-layered concrete sandwich elements with a thin layer of core insulation.

III KfW: Kreditanstalt für Wiederaufbau (Credit Institute for Reconstruction) is a bank that was established in Germany to administer public funding and low-interest finance for building projects.

IV   Düsseldorf Building Department had instructed the architects Bernhard Pfau, Hanns Junghans, and Paul Schneider-Esleben to develop modern educational buildings.

V   Asbestos was found primarily in the fire doors and fire dampers and in special asbestos cement components in the school. products containing SMF that were manufactured prior to 1966 and which are therefore classed as carcinogenic according to the Hazardous Substances Ordinance (Gefahrstoffverordnung = GefStoffV), were discovered in the insulation material of electric wiring and in filler compounds. In addition, PCB was found in paint coatings, curtains, and floor coverings.

VI   In this case: the PCB Ordinance in the federal state of North Rhine-Westphalia.

VII This change is also evident in the building's forecast heating energy requirement: For 2010 the requirement is estimated at 1,700 MWh/a, which compares to an average heating energy demand of 2,100 MWh/a in the years 2000 to 2009.

UNIVERSITY OF STUTTGART'S K II BUILDING

STUTTGART (DE)
Educational —pp. 58, 95, 98, 152

— Heinle, Wischer und Partner

1 Constructed between 1960 and 1965, the university buildings K I and K II are typical examples of the Stuttgart school style.
2,3 The juxtaposition of sections with three lower and two higher stories respectively turns the building into a spatial wonder and its circulation core into an architectural mountain landscape comprising steps, bridges, and galleries.
4 The foyer of the K II university building

5

6

7

8

Kollegiengebäude II (K II) is one of the most characteristic examples of buildings in the second so-called Stuttgart School style. Together with its neighboring twin (K I), its compact and orthogonal large-scale form is amongst the most distinctive buildings of the Technical University. The building, containing libraries, lecture halls, and seminar rooms, is used today primarily by the institutes of the Faculty of Humanities and the Faculty of Management, Economics, and Social Sciences. K I, built in 1960, and its nearly identical descendant K II, built in 1965, were designed by university professors Rolf Gutbier, Günter Wilhelm, and Curt Siegel. They were erected on the inner-city campus as replacements for buildings destroyed in the war.

EXPOSED CONCRETE: SOMETIMES ROUGH, SOMETIMES SMOOTH

REFURBISHMENT OF THE UNIVERSITY OF STUTTGART'S K II BUILDING
HEINLE, WISCHER UND PARTNER, COLOGNE

Four decades later, the practice Heinle, Wischer and Partner—whose founders had participated in the original planning of K II—were commissioned with its complete refurbishment. The purist clarity of the K II building pleased the architects so well that, in the course of the spatial reorganization, they only intervened in its structure and its design when absolutely unavoidable. The visible parts of the structure were repaired, cleaned and, where needed, replaced in the same form. The refurbishment comprised the replacement of building services systems, the roof and the north facade, suspended ceilings, and floor coverings. Additionally, a large amount of toxic materials had to be removed. Departmental libraries, lecture halls, institute areas, and the cafeteria were restructured in accordance with their changed use: The library's floor area was increased, for instance, and the lecture halls were upgraded with regard to ventilation and media technology. Even the mezzanines are now accessible for people with disabilities. Outwardly unchanged, the K II building has been transformed through targeted interventions "behind the scenes," into a building that meets the newest standards for energy and technology. FPJ

PROJECT DATA

Client State of Baden-Wuerttemberg/ Stuttgart regional tax office, represented by the university building department
Construction costs EUR 16.1 M
GFA 26,886 m²
Gross floor area 26,886 m²
Usable floor area 11,586 m²
Primary usable area 10,555 m²
Completion 9/2009
Project management Winfried Schmidbauer, Monika Horn
Location Keplerstraße 1, D–70174 Stuttgart

Year of construction 1965

1000 ———————————— 2010

Conversion 2009

2000 ———————————— 2010

Construction costs   **Price per m²**
10.1 M €   1,390 €

15,000
10,000
5,000
0

Thanks to an ingenious coupling of multi-floor groups, each comprising three low stories for the institutes on one side, together with two higher stories for seminar rooms on the other, Gutbier and his partners were able to optimize the building's usable floor area. Thus, the high-rise building has fifteen stories with office use on its south side, and ten stories accommodating the libraries and rooms for teaching on its north side. Accordingly, the floor-to-floor height varies from 2.90 m in the offices on the south side to 4.35 m in the lecture halls on the north side. Elevators, technical rooms, and sanitary facilities, as well as the stairs, are located in the partly open core zones. The open corridors with their single-flight stairs each give access to three levels, which also have a spatial presence due to the open configuration. This type of circulation transforms the "servant" building core into an architecturally delightful, three-dimensional stage.

In accordance with the maxim of the Stuttgart School, as derived from classic Modernism—visibility of the structural principle, functionality, and restriction to a few basic materials—the appearance of the building is one of exposed concrete, masonry, glass, and wood. The designers were especially fond of exposed concrete, which characterizes the building inside and out: sometimes executed with a rough texture, sometimes with a lively one, and sometimes using smooth formwork, but always exposed and unplastered. "Bare visible concrete, without plaster and not a single brushstroke on the load-bearing and space-forming frames and wall surfaces, . . . the fewest forms and colors for the few additional elements such as window and door frames, etc. Built for the scrutinizing eye of future civil engineers and architects. . . ." That is how the architect Günter Wilhelm described the guiding principles of the design.

During rehabilitation work

TYPICAL GROUND FLOOR PLAN

10 m

CROSS SECTION THROUGH THE BUILDING

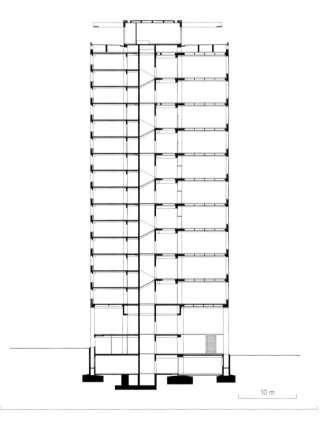

10 m

INSTITUTE OF VEGETABLE AND ORNAMENTAL CROPS

GROSSBEEREN (DE)
Research

— Numrich Albrecht Klumpp Architekten

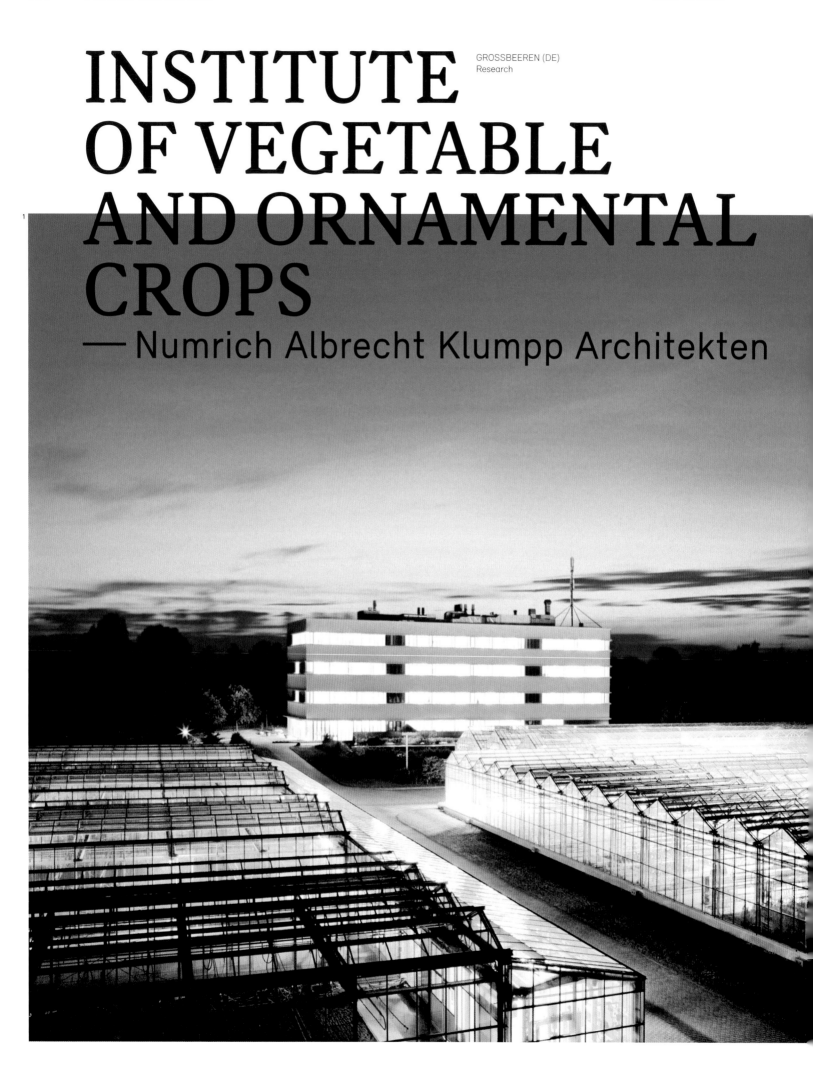

1   The Institute of Vegetable and
Ornamental Crops: the vegetables of
the future are bred in greenhouses
and research is conducted in the
white institute building.
2–4 Cafeteria and staff kitchen on
the ground floor

THE WHITE BAND

RENOVATION OF A LABORATORY BUILDING IN GROSSBEEREN, BRANDENBURG

NUMRICH ALBRECHT
KLUMPP ARCHITEKTEN, BERLIN

The Institute of Vegetable and Ornamental Crops (IGZ) in Groß-beeren still enjoys the outstanding reputation that it built up in communist times. Now, as then, plants are analyzed and cross-bred so as to develop new varieties in the greenhouses of the research facility south of Berlin. The institute's laboratories are located in a four-story, prefab-ricated East German building from the 1970s (type SKBS 75). The steel-framed structure and its gridded, exposed-aggregate concrete facade had not only become unsightly, but rehabilitation in functional and mechanical terms was also overdue. It seemed advisable to replace the prefab with a new building, but that would have exceeded the budget constraints and also necessitated temporary quarters for the laboratories. Since the space this would require was not available on the institute's campus, the architects and the institute's management decided to refurbish the building during ongoing use: The long rectangular building was divided down the middle with a bulkhead. Research continued on one side while the building was renovated on the other.

Numrich Albrecht Klumpp Architekten retained the building's structure as well as the stairways. Although built according to a standardized, modular system, reinforced panels of varying thickness were encountered throughout the building, so the reuse of the load-bearing elements required a structural analysis. Among the most significant interventions, in addition to installing an elevator, was shifting the main corridor, originally located in the middle, to create an asymmetrical, one third/two thirds longitudinal division. The staff offices are located on the narrow side and the laboratories on the wider side. By inserting glazing between the offices and the corridor, the long hallway also receives natural lighting.

The renovation has transformed a dismal standardized building, which though admittedly still austere, is now expressive and distinctive, thanks to the powerful plasticity of its snow-white spandrel bands. These alternate with three rows of ribbon windows. Beneath the smooth stucco surface, the building has received new thermal insulation all around.

Sometimes it just takes a small step to go from a featureless, functional building to a well-designed one. In the Großbeer-en laboratory building, this leap essentially has to do with the molding below the windows, because it effectively emphasizes the predominant motif of the white stucco bands and lends the facade three-dimensional depth. During the day, these bands shine brightly between the darker ribbon windows, whereas at night, this relationship is reversed. From a distance, the continuous window moldings appear to be solid, but they are in fact hollow: sheet metal enclosures for the sun shading. "We originally wanted to construct the moldings of solid concrete," explains the architect Werner Albrecht. But this solution, as the bidding showed, would have cost four times the price of the metal housing that was ultimately built.

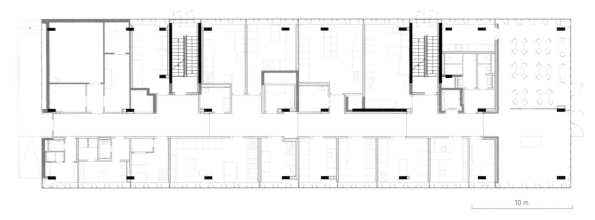

Condition before renovation

Inside, the architects have placed colorful accents where the corridors are widened. The cellular layout of the floor plans has been maintained at the request of the client, but next to the entrance on the ground floor, the building envelope has been opened for a dining room with an adjacent kitchenette. When the scientists choose to leave their microscopes or computer screens, they meet here—the room is very popular. FPJ

PROJECT DATA

Client Institut für Gemüse- und Zier-pflanzenbau Großbeeren/Erfurt e.V.
Construction costs EUR 4 M (cost groups 300–400)
Usable floor area 2,020 m²
Gross volume 10,800 m³
Feature Laboratory renovation while in use
Completion 3/2008
Project management/team Werner Albrecht, Grant Kelly
Location Theodor-Echtermeyer-Weg 1, D–14979 Großbeeren

Year of construction 1988

| 1000 | 2010 |

Conversion 2008

| 2000 | 2010 |

Construction costs 4 M € **Price per m²** 1,980 €

15,000
10,000
5,000
0

GROUND FLOOR PLAN

10 m

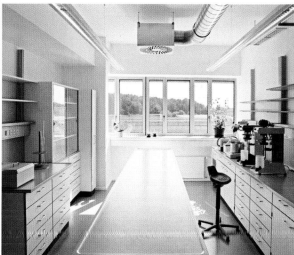

5   The new entrance area
6   View of a laboratory
7, 8   Strong, warm colors make the
stairs and corridors stand out.

SIESMAYER- STRASSE OFFICE BUILDING

FRANKFURT AM MAIN (DE)
Office — pp. 72, 76, 132

— schneider+schumacher

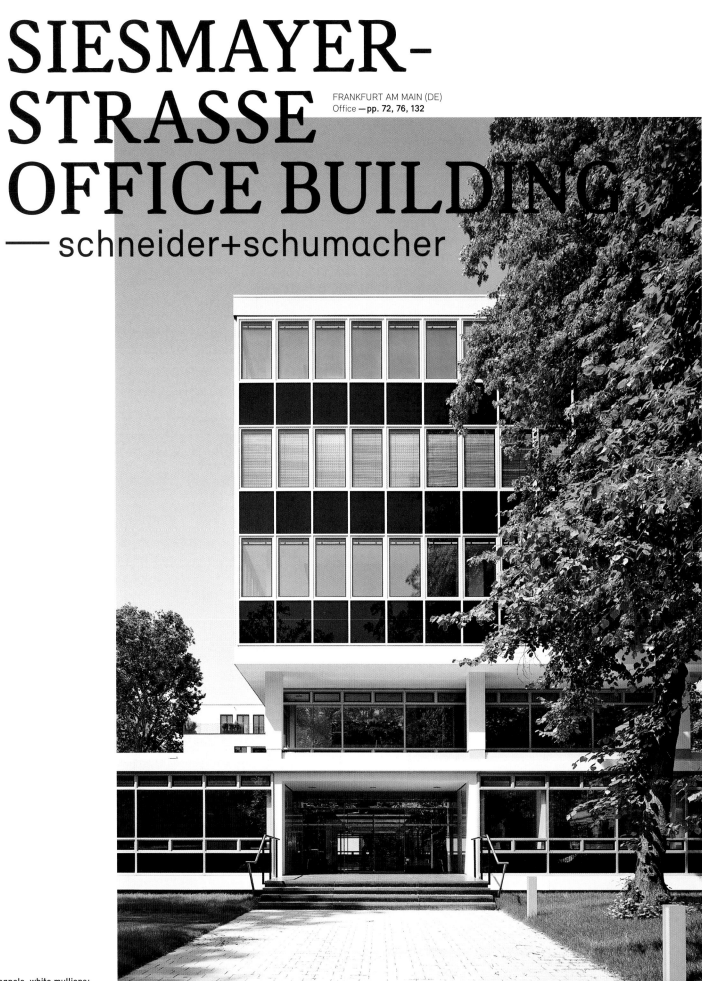

1   Black panels, white mullions: entrance and main facade after refurbishment

"What you have inherited from your forefathers, you must first win for yourself if you are to possess it." [Goethe]

The curtain wall facade—developed as early as 1920—finally gained widespread acceptance in the 1950s with the construction of office buildings, especially in bustling Frankfurt am Main, which emerged from the dust of the wartime rubble faster than other cities. In 1955, the large American architectural practice Skidmore, Owings & Merrill LLP designed an office building, in Siesmayerstraße on a prime location in Frankfurt's Westend area, that can rank as a paradigm of the International Style. In fact, the cool elegance and clearly articulated lines of the architecture were a stylistic re-import, since the major impetus for mainstream post-war Modernism had been given by emigrants from Europe, such as Gropius and Mies van der Rohe. Yet their American contemporaries had long since developed their own characteristic style, with a sober but sophisticated aesthetic of the grid. The client for the five-story building used as the consulate general was the United States of America. In parallel with the complex in Siesmayerstraße, SOM designed three additional consulates—in Bremen, Düsseldorf, and Stuttgart—using the same "kit of parts"—in each case, however, with a different form. Because of its prototypical character for early post-war Modernism, the building was placed on the heritage list of landmark buildings in 1986. After the Americans moved out in 2005, the building was sold. The new owner wanted to refurbish the building in consultation with the landmark preservation authorities and taking into account current norms for thermal insulation, sound insulation, and fire protection. The interior fittings exhibited noticeable signs of wear and tear, were completely destroyed in places, and needed a new rendition. It quickly became apparent that this was less a matter of conserving and renovating the existing fabric than of preserving the spirit of the architecture.

AS ELEGANT AS EVER

RENOVATION OF THE FORMER US CONSULATE GENERAL IN FRANKFURT
SCHNEIDER+SCHUMACHER, FRANKFURT

The "swing"—the buoyant lightness of the 1950s—ought to resound again. Thus the renovation was carried out with the goal of enabling the most flexible room layout possible for general office use on all the floors. To this end, units comprising 400 sq m were created. On the ground floor, multiple units can be combined in compliance with building regulations. The original reinforced concrete frame, with beams and coffered ceilings, was in good condition and was therefore retained to a great extent. Only the ceiling slab above the ground floor was partially dismantled and rebuilt, since it exhibited significant structural deficiencies. The elevator core and the existing stairs were demolished, rebuilt in another location, and redesigned in keeping with the spirit of the building. In place of the previous crawl space beneath the building, an underground parking garage was built.

SECTION THROUGH THE RENOVATED BUILDING

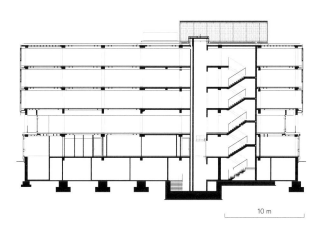

10 m

Now, after its refurbishment, the facade's outer appearance corresponds precisely with that of the old one, both in the material and in the profile sizes, which the architects have meticulously reconstructed. In terms of technical performance, it conforms to present-day demands for sustainability. On the ground floor and the first above-grade floor, the facade consists of a post-and-rail system with outward-opening sashes. The facade grid here measures 2 m by 1.89 m. External sun shading has been added on these stories.

On the second to fourth above-ground floors, with their narrower grid of merely one meter in width, the architects have also replaced the post-and-rail-facade, but used tilt-and-turn double-sash windows. The sun shading is located in the interstitial space between the outer single glazing and the inner insulated glazing. Every room now has direct natural lighting and ventilation. The energy-intensive split units for cooling have been replaced with efficient radiant cooling panels set into the ceiling. A water basin in the courtyard provides evaporative cooling, further improving the quality of the air in the surrounding offices. Today, a large American law firm is the building's main tenant.

The old consulate building is an example of how the original architectural intentions of a building from the 1950s can be carried over authentically into the present without neglecting energy considerations. FPJ

PROJECT DATA
Client G & P Grundstücksentwicklungs GmbH & Co. Siesmayerstraße KG
Construction costs approx. EUR 5 M
Lot size 7,829 m²
GFA 4,050 m² above ground, 1,995 m² below ground
Usable floor area approx. 3,200 m² (above ground)
Feature Reinforced concrete frame (existing building from 1955) with post-and-rail facade restored according to requirements of landmark preservation agency
Completion 12/2007
Project and construction management Michael Schumacher, Kai Otto, Peter Mudrony
Location Siesmayerstraße 21, D–60323 Frankfurt am Main

Year of construction 1955

1000   2010

Conversion 2005–2007

2000   2010

Construction costs   **Price per m²**
5 M €   310 €

15,000
10,000
5,000
0

The building before refurbishment, rear view

2   The redesigned internal courtyard appears almost Japanese in
terms of its clarity and tranquility.
3   The vertical spacing of the curtain wall elements is 1 m.
4   The windows have retained
their original appearance after
refurbishment.

TEMPERED TRANSPARENCY
— Renovating Modern Movement Facades in Line with Landmark Preservation Principles

The transparent building envelopes that were created in the first two decades of the twentieth century were an expression of a new understanding of architecture, and of the innovative building concepts associated with it. However, it was not until after the Second World War that they became the predominant style for public buildings and, above all, commercial developments.

UTA POTTGIESSER/JULIA KIRCH

In post-war buildings in the Modernist style we can find significant improvements in construction details, in the form of linked casements and double glazing, compared to the buildings of Walter Gropius et al. These were continually improved as legal requirements for energy conservation in buildings became stricter. However, even the facades of the 1980s are a long way behind today's standards. Refurbishment concepts for such building envelopes always focus on the question of how these style-defining components can be retained, revitalized or modified as part of an upgraded energy conservation concept.[1]

DEVELOPMENT OF CONSTRUCTION AND ENERGY-CONSERVATION STANDARDS
—

The facade details of office and administrative buildings of the early Modern Movement consist of simple steel profile angles with puttied single glazing such as that of the Bauhaus building in Dessau, which was completed in 1926. The construction methods and materials developed at the time were an expression of the desire for an architectural style of "dematerialized" glass buildings. However, these also raise fundamental questions in relation to the demand for improved working and living conditions. Along with the advance of technology, user requirements called for high-quality habitable interiors.

Whereas the profiles used towards the end of the nineteenth century still resembled those used in timber construction, they were simplified in their geometry at the beginning of the twentieth century. The size of glass panes was rapidly increased in the course of the 1920s, as was the overall proportion of glazing. As a result, buildings were subject to overheating during the summer months and considerable heat loss in winter. The Modernist avant-garde of the was little concerned about energy conservation. An exception is the *mur neutralisant* concept, which was developed by Le Corbusier in 1929: It means "neutralizing wall" and was created as part of the design for the

---

1   [p. 124] The Bauhaus building in Dessau: The facade construction of the studio building by Walter Gropius, built in 1926, with small-format single glazing.

2   As built, the double-skin facade of the Centrosojus building in Moscow, begun in 1929 to a design by Le Corbusier, was not ventilated.

Centrosojus office building in Moscow. It took the form of a mechanically-ventilated facade cavity, which was meant to balance the extreme annual temperature fluctuations characteristic of Moscow.[II] However, this cavity wall concept was not put into practice until the second half of the twentieth century.[III]

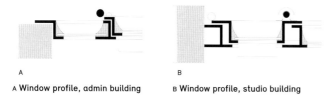

DRAWINGS SHOWING THE SIMPLE GEOMETRY OF THE PROFILES USED ON THE BAUHAUS BUILDING
The putty used on the glass panes also serves as a seal.

A   Window profile, admin building   B   Window profile, studio building

The first linked casement windows appeared on the market at the beginning of the twentieth century. At the beginning, they consisted of doubled-up steel profiles with a single glazing panel each, but these gave way increasingly in the 1930s to systems consisting of two separate casement frames. During this period, the hard-edged geometry of L, Z, T, and H shaped steel profiles was refined to include tapered contact faces, which

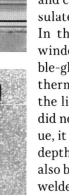

improved the air-tightness of the frame construction. Although this resulted in a marginal improvement of the U-value of the glazing elements, there was not really a solution to the problem of thermal bridging of steel profiles, and component joints, as well as of poorly insulated walls and parapets, until the 1950s. In the 1950s and 1960s, linked casement windows were increasingly replaced by double-glazed windows in single frames without thermal breaks. Although, in comparison to the linked casement window, double glazing did not achieve an improvement in the U-value, it did result in a reduction of construction depth. Special forms of double glazing should also be mentioned, which featured soldered or welded edge details, the latter with a gap between the panes of just a few millimeters.[IV]

The steel glass facades of the 1950s were almost all constructed using manual methods, whereas from 1963 onwards these systems were increasingly replaced by industrially-produced aluminum facade systems. The early aluminum constructions of the 1960s also lacked thermal breaks and were therefore insufficient with regard to thermal insulation and thermal bridging at the connection points, which led to frequent moisture problems. The breakthrough—and hence an important improvement in buildings' physical

performance—was achieved with the introduction of thermally broken window and facade profiles at the beginning of the 1970s. In Germany they were used for the first time in buildings such as the BASF building of 1957 in Ludwigshafen, by HPP Hentrich, Petschnigg & Partner, and the Nationalhaus of 1964 in Frankfurt, by Max Meid. This development was also a reaction to the oil crisis in 1973, which had awakened the public's awareness of the need to use resources efficiently. This led to the passing of the first Energy Conservation Law (EnEG) in 1976 and the first Thermal Insulation Ordinance (WSchVO) in 1977.

Since the mid-1970s, the thermal insulation of the envelopes of buildings has undergone continuous improvement through the optimization of glazing and framing components. From the 1960s it had become necessary to compensate for disadvantages such as overheating and heat loss, resulting from the high proportion of glazing in facades, by the mechanical air conditioning of the interior spaces. Frequently there was neither natural ventilation through opening sashes nor an effective

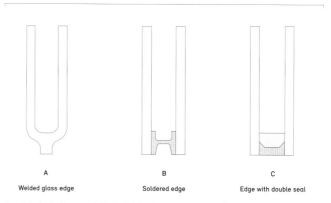

A   Welded glass edge   B   Soldered edge   C   Edge with double seal

Insulated glazing was initially fabricated with soldered (e.g. Thermopane), edge-welded (e.g. GADO, SEDO) and adhered (e.g. CUDO) edges.
Since roughly 1960, the principle of an adhered edge with a double peripheral seal has prevailed.

exterior solar screening system. Although the new air conditioning technology was used to compensate for these deficits in the buildings, the use of this technology also led to increased energy consumption, while occupants experienced problems relating to health and general well-being, a phenomenon that was soon given the name Sick Building Syndrome (SBS).[V] It was not until the 1990s, with the development of climate concepts for buildings, that the holistic assessment of energy efficiency and energy balance in buildings that is in use today could be effectively promoted, becoming established finally with the introduction of the 2002 Energy Conservation Directive.

The architectural quality of post-war Modernism—in particular the period after 1960—was, for a long time,

3   Today, the Centrosojus in Moscow is marred by building defects and post-war modifications.

4   GADOGlas, an edge-welded glazing product, was used in 1958 by Le Corbusier for the Unité d'habitation in Berlin

DEVELOPMENT OF THE WINDOW PROFILE IN THE TWENTIETH CENTURY
Constructions and U-Values of Steel and Aluminum Window Profiles from 1905 to 2005

PROFILE/SYSTEM	STEEL		PROFILE/SYSTEM	STEEL	ALUMINUM

1905

First linked-casement window produced from industrial sections with leaf-sprung seals

U-value 3.6

1963

Increasingly, steel profiles are replaced by aluminum profiles.

U-value no data   **U-value** no data

1915

Single-glazed windows from drawn hollow sections; contact faces are lined with rubber.

U-value 5.9

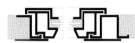

1971

Aluminum profiles are further improved, e.g. by increasing their rigidity.

U-value no data   **U-value** no data

1929

Linked-casement window made from special sections; the contact faces of the sections are still parallel.

U-value 3.6

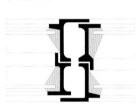

1980

Steel profiles are offered with foam core insulation.

U-value no data   **U-value** 3.5

1931

Linked-casement window using chamfered special sections with three contact faces and that are no longer fitted with putty, but held with glazing bars.

U-value 3.6

1990

Steel profiles are fitted with thermal breaks as standard.

U-value no data   **U-value** 2.6

1953

Glass panes are fixed with metal glazing bars.

U-value 3.6

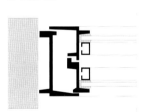

2000

Significant reduction in U-value by forming cavities in the profiles (steel and aluminum)

U-value no data   **U-value** 1.8

1958

First double-glazing in single-sash windows

U-value no data

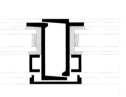

2005

Further reduction in U-value by reducing thermal bridging

U-value no data   **U-value** 1.4

DEVELOPMENT OF THE U-VALUES OF WINDOWS AND FACADES SINCE 1960
Through Improved Insulated Glazing and Frame Construction

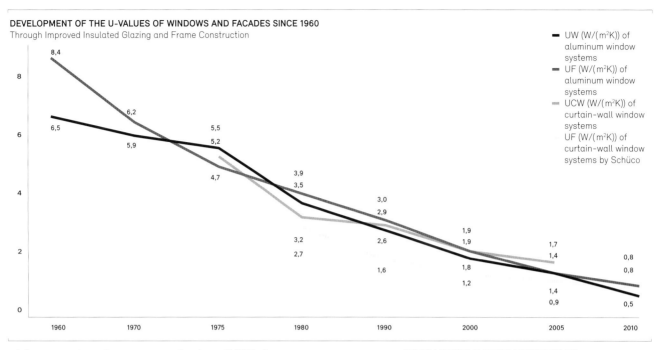

either questioned or not perceived at all.[VI] Today numerous buildings from the 1950s and 1960s are protected as historic buildings. Current discussions about the quality of architecture worthy of preservation revolve mainly around buildings from the late 1960s and early 1970s.[VII] Some of the buildings of that period are particularly at risk, owing to today's more stringent energy conservation requirements.

REVITALIZATION STRATEGIES
FOR MODERN
FACADE CONSTRUCTIONS
—

For the purpose of revitalizing modern facade constructions we need to distinguish primarily between buildings with landmark preservation and those without, between buildings with different functions and in different states of repair. In historic buildings with landmark status, the first priority is to retain the original appearance. In respect of building categories, office buildings have different sets of requirements than do residential developments, or buildings for cultural and industrial uses. The condition of the respective building (or its components) plays an important role regarding the necessary scope of the revitalization. After all, building owners and users are mostly concerned about upgrading interior comfort levels, thermal and acoustic insulation, and adapting workplaces to today's requirements. With respect to landmark preservation, it is possible to distinguish three fundamental refurbishment strategies for facade constructions dating from the twentieth century:[VIII]

RETAINING THE EXISTING
FACADE CONSTRUCTION
—

This is the classic approach to landmark preservation. In this case only certain elements are modified while largely preserving their original appearance, for example:
- preserving and repairing profiles,
- reconstructing the original profiles,
- replacing the original glazing with K-Glass™ or insulating glazing,
- replacing or improving sealing profiles.

With these measures it is usually only possible to achieve minor improvements in thermal insulation. However, overall energy consumption can be reduced to a certain extent by installing supporting ventilation and air conditioning systems.

SUPPLEMENTING THE EXISTING
FACADE CONSTRUCTION
—

This is also a common approach in landmark preservation. Here the existing facade is retained as outer layer. Possible variations are:
- installing an additional interior glazing or insulation layer to fenestration or walls,
- installing an additional exterior glazing or insulation layer to fenestration or walls, or as facade or fenestration bands which have a significant impact on the exterior design of the building.

These supplementary measures help to achieve current standards and make significant improvements in thermal insulation and interior comfort. However, where it is not possible to adapt ventilation and screening systems fully to the new uses, the potential for reducing overall energy consumption cannot be fully utilized.

REPLACING THE EXISTING FACADE CONSTRUCTION
—

This is also a common approach in landmark preservation, making it is possible to recreate the original appearance. In other buildings, this option is the most common one and involves either:
- installing a single-skin facade, or
- installing a double-skin facade.

The current standards of facade systems lead to a significant improvement of thermal insulation in buildings and interior comfort levels and, through the use of new, adapted ventilation and screening systems, also achieve a significant reduction in the overall energy consumption. In Europe the energy conservation standard of facade constructions has been continuously improved over recent decades through ongoing development of refurbishment methods. However, since about two thirds of office buildings in Central Europe were built before 1978, there is still a large proportion of existing buildings which await an upgrade in construction and energy efficiency.[IX] The strategic approaches above are illustrated in some selected examples for the revitalization of modern facade constructions.

PRESERVATION. HAUS HARDENBERG, BERLIN (PAUL SCHWEBES, 1956)
—

With its clear structural elements and fenestration bands, Haus Hardenberg by Paul Schwebes follows on from the new buildings of the 1920s and is considered one of the most beautiful German buildings of the 1950s. The building with its trapeze-shaped layout impresses with its dynamically rounded corner solution and filigree projecting roof.

The main facade features a classical division into three zones, with a plinth, main, and receding attic floor zone, which are structured horizontally with protruding concrete decks. The five upper floors are characterized by fenestration bands consisting of story-height elements with large fixed glazing panes at the center and small opening sashes on both sides, reminiscent of the "Chicago window." The window units are subdivided by vertical brass-colored steel profiles and the slender steel frames are painted black on the outside and white on the inside. The window elements are designed as linked casement windows and are characterized by small gaps between the window and thermal insulation in the facade. The slender geometry of the profiles, together with the selected materials and colors—black, white, and brass—give the building its elegant appearance.

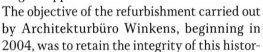
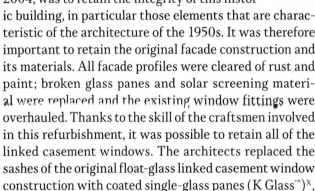

The objective of the refurbishment carried out by Architekturbüro Winkens, beginning in 2004, was to retain the integrity of this historic building, in particular those elements that are characteristic of the architecture of the 1950s. It was therefore important to retain the original facade construction and its materials. All facade profiles were cleared of rust and paint; broken glass panes and solar screening material were replaced and the existing window fittings were overhauled. Thanks to the skill of the craftsmen involved in this refurbishment, it was possible to retain all of the linked casement windows. The architects replaced the sashes of the original float-glass linked casement window construction with coated single-glass panes (K Glass™)[X],

DIAGRAM OF COMMON REFURBISHMENT STRATEGIES FOR FACADE CONSTRUCTIONS

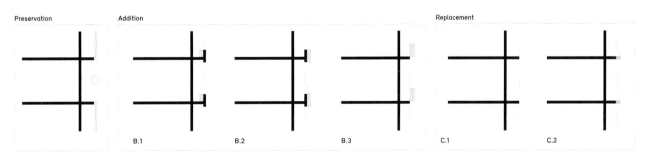

Preservation   Addition   Replacement

B.1   B.2   B.3   C.1   C.2

B.1 Additional inner glazing or insulation layer

B.2 Additional outer glazing or insulation layer

B.3 Continuous plane

C.1 Installation of a single facade
C.2 Installation of a double facade

5   [p. 124] **Haus Hardenberg in Berlin**, built in 1956 to a design by Paul Schwebes. State following the refurbishment of the original facade in 2004

6   The fenestration bands are structured by vertical steel profiles. The slender profiles, together with the selected colors—black, white, and brass—give Haus Hardenberg its elegant appearance.

so that it was possible to reduce the U-value of the linked casement window construction. In spite of this improvement to the glazing system, the lack of thermal breaks in the steel profiles meant that it was not possible to achieve today's requirements for thermal and sound insulation. For this reason, the shops and offices in the building are mechanically ventilated and air conditioned via ceiling units. Other parts of the building services were adapted to current safety requirements.

The strategy of retaining as much of the existing building as possible was also pursued with great success by Brenne Architekten in their refurbishment of the ADGB school in Bernau, which was constructed in 1928 by Hannes Meyer [p. 184]. The architects succeeded in retaining and overhauling most of the facade construction. In a few cases, frames were rebuilt using new steel profiles of identical dimensions.

ADDITION: BEROLINAHAUS, BERLIN (PETER BEHRENS, 1932)
—

The Berolinahaus building was completed in 1932 to a design by Peter Behrens; together with the Alexanderhaus building, it forms a gateway to the redesigned Alexanderplatz. These buildings are some of the best known examples of the New Objectivity (Neue Sachlichkeit) style in Berlin. The project had been the subject of a competition organized by the Berlin city authority (Magistrat) in which all notable representatives of the New Building (Neues Bauen) style took part. Behrens's buildings feature clearly structured, perforated facades with symmetrical window grids, as well as a remarkable concrete frame construction. The front elevation of this building is dominated by an imposing vertical light-well feature in frosted glass, which extends well beyond the roof line.

HAUS HARDENBERG IN BERLIN Detail of the window construction

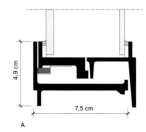

A Original construction
B Execution with original profiles, but with new inner glazing

(K Glass™) and new seals, as rehabilitated by Winkens Architects in 2004

The upper floors feature large square window elements, which correspond to the construction grid of the load-bearing reinforced concrete frame. Whereas we must assume that the window and door frames on the first floor were made of brass profiles, the square window elements of the upper floors were manufactured using rolled steel profiles. These elements were subdivided into square glazing panels with pivot sashes, which could be reversed fully. Originally, the glass panes in these externally-mounted single-glazed windows were installed in a putty bed. In later years, a kind of double glazing was introduced by fitting secondary panes. In addition, internal wooden windows were fitted during the post-war period.

In 2005, the architects nps tchoban voss, with Sergei Tchoban and project partner Philipp Bauer, were commissioned to carry out the refurbishment of the facade in accordance with landmark preservation principles. The repair of the natural stone surfaces presented a considerable problem. The facade, which during the war had suffered damage from bullets and shrapnel, had been repaired during the communist years, using a jetcrete application. The sprayed-on concrete had bonded with the shell marl substrate and formed a glass-like compound which could not be removed without causing further damage. Since the original material could no longer be obtained, a similar stone from the Elm area was selected, which has a somewhat different, coarser texture. To improve the energy balance of the building, calcium-silicate boards were installed as internal insulation.

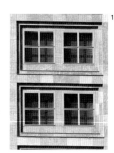

As part of the refurbishment, the 360 windows of the upper floors were fitted with new internal units, constructed of thermally broken aluminum profiles and double glazing, which were installed in the same plane as the thermal insulation in the walls. This means that the external steel windows have become the outer parts of a linked casement window, which is back-ventilated via open joints. Except for one axis of the building, which was fitted with overhauled original windows, the architects replaced all historical external windows with windows made of specially manufactured, extruded aluminum profiles without thermal breaks. In this way it was possible to retain the external division of the facade, as well as the face widths of the original window profiles. As the original materials and coloring of the shop windows on the first floor and the oriel band along the second floor were not known, the architects opted

7   [p. 125] See caption 6, p. 123

8   The steel linked-casement windows of Haus Hardenberg are fitted with single glazing in the inner and outer sashes.

9,10 [p. 125] Prior to the refurbishment, the facade of Berolinahaus, which was constructed by Peter Behrens in 1932, was marred by unsympathetic post-war intervention.

11   [p. 125] The facade after refurbishment to reinstate the original design

for golden anodized aluminum profiles, which are similar to those of the slender window frames. The inside is fitted with a thermally separated window construction with double glazing. Instead of the original frosted opal glass, white safety glass was installed, which was backed with white enameling, so that the effect of the original glazing could be recaptured.[XI]

REPLACEMENT: OFFICE TOWER OF THE EUROPA CENTER, BERLIN (HENTRICH-PETSCHNIGG & PARTNER, 1964)

—

The Europa Center on Breitscheidplatz in Berlin dominates the center of what was once West Berlin. The building complex, which consists of three buildings with a shop-in-shop mall on the first floor and a high-rise office tower on the upper floors, was completed by architects HPP Hentrich-Petschnigg & Partner in 1964. Together with the New Memorial Church by Egon Eiermann, which incorporated the ruin of the Kaiser Wilhelm Memorial Church, the Europa Center became a symbol of reconstruction and a landmark of the inner city of West Berlin. The low-rise parts of the building and the office tower are clad in an aluminum post-and-rail construction, using a structuralist design concept to create a building in the International style, which had developed from classic Modernism. A concrete frame with two parallel rows of pillars supports the concrete decks, covering a footprint of 47.30 × 17.30 m. The pillars are recessed from the facade and front face of the decks by about one meter, so that the curtain-wall facade envelops the building as a homogeneous skin. The office tower was designed with a single-skin facade based on a 1.875 m modular grid. Large fixed glazing units alternate vertically with opaque spandrel panels between the continuous projecting posts of the facade. The interior is subdivided into individual offices. The two office units, with their artificially lit corridor zones, are connected to a central services core.

UTA POTTGIESSER

—

Prof. Dr.-Ing. Uta Pottgiesser has worked as an architect in Berlin and is currently professor of building construction and building materials at the Hochschule Ostwestfalen-Lippe-University of Applied Sciences in Detmold. In 2002 she completed her doctorate on "Multi-layered glass constructions" at the TU Dresden, Institute for Building Construction. At the Detmold School of Architecture and Interior Design she is the spokesperson for the research project "ConstructionLab" and co-founder of the master's program "International Facade Design and Construction." She is the author of various textbooks and specialist works in the field of building construction and facades, including "Fassadenschichtungen-Glas."

JULIA KIRCH

—

Dipl.-Ing. (FH) Julia Kirch is a research assistant working on the research project "ConstructionLab" at the Hochschule Ostwestfalen-Lippe-University of Applied Sciences. She previously studied interior design and then architecture while working in various architectural offices. While employed at the Detmold School of Architecture and Interior Design, she contributed to diverse publications in the fields of building construction and facades.

www.hs-owl.de/fb1
www.constructionlab.de

The conversion of the existing office tower into a modern office building was also carried out by architects HPP Hentrich-Petschnigg & Partner in 2001. One objective was the refurbishment of the aluminum curtain wall facade, the appearance of which is typical of the urban center of the former West Berlin. Another objective was the transformation of the interior to create a modern office landscape with up-to-date furnishings and corresponding comfort levels. The existing curtain wall facade was replaced with an aluminum double-skin facade with linked casement window elements. With the help of this new double-skin facade and an externally fitted single-glazing system, it was possible to replicate the slender external appearance of the original facade with its narrow, vertically arranged profiles. The outer glazing system effectively conceals the wider profile sections required for the double-glazing fitted on the inside. While the inner layer functions as thermal separation, the outer skin and the narrow interstice protect the building from the effects of weather such as solar radiation,

wind, and rain. In contrast to the original design, the new facade allows natural ventilation of some of the rooms via ventilation slots in the external single-glazing skin, above and below the spandrel panels. The inner window elements have opening sashes. This construction significantly cut heat transmission losses through the building envelope, as well as noticeably reducing operating costs and the energy used for air conditioning.

It also meant that much less space was needed for the installation of services equipment. Most of the space gained was utilized for offices. Overall, the new layout design provides greater flexibility in the use of the building. The new layout provides classic cellular offices, as well as group, combination, and open-plan offices. In addition, all building services installations, such as heating and ventilation ducts and electric and data cabling, are placed in to a services duct which runs

12	[p. 128] In the refurbishment by HPP in 2002, the single-skin facade of the tower block was replaced by a double-skin facade system which appears almost identical from the outside.

13	[p. 128] Section of the facade: Newly installed solar screening elements can be seen behind the outer skin.

all along the periphery. This meant that it was possible to use transparent partition walls in many instances, which substantially enhanced the natural lighting of the office zones.

A similar strategy was followed by architects schneider+schumacher in their revitalization project

for an office building in Siesmayerstraße in Frankfurt. The building, designed by architects Skidmore, Owings & Merrill (SOM), had been constructed in 1955 to accommodate a US General Consulate. The original steel-frame building had been constructed without bracing walls, and after refurbishment, retained its characteristic division into a flat plinth, a set-back mezzanine floor and a slender tower. So that not too much space would be lost owing to the thicker facade construction, which was required to meet energy conservation standards, an additional grid section was added to the length and width of the high-rise building. The curtain wall facade, which at the time was a novel feature, is a metal construction. As a key example of its type, it enjoys landmark preservation status and has been carefully renovated, taking into account the current regulations on thermal and sound insulation and fire safety. In spite of the technical upgrade, the exterior appearance of the facade remains unchanged: On the first and second floors, the original single-skin double-glazing consisting of top-hung and bottom-hung sashes was reinstated and supplemented with external solar screening. On the upper floors, the windows comprise an inner double-glazing layer and an external glass pane with solar screening elements in the interstice. Originally the building was designed without solar screening. The lack of any shading led to extensive overheating of the interior and it was therefore decided to equip the whole building with solar screening elements as part of the refurbishment. The architects were concerned to match the color scheme of the facade and its profiles as closely as possible to the original. In order to match the champagne color of the profiles accurately, it was necessary to carry out many color tests and inspections.

REVITALIZATION PROJECTS OUTSIDE EUROPE
—

Buildings in the International or Modern Movement style in other European countries and outside Europe have

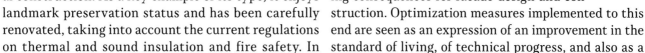

experienced similar structural and physical problems. Often these defects are compounded by questionable construction standards and low quality building materials. Since the 1930s, facade designs and building concepts developed for Europe and temperate climatic zones have often been exported to other continents without sufficient adaptation to local climatic conditions. For many countries in Central and South America, as well as Asia, suitable refurbishment concepts have yet to be developed. Discussions with scientists, as well as practicing architects and engineers, in countries such as Brazil indicate that the comfort levels and energy efficiency required by occupants in these countries, too, are greater than they used to be—with the corresponding consequences for facade design and construction. Optimization measures implemented to this end are seen as an expression of an improvement in the standard of living, of technical progress, and also as a necessity to address climate change issues. In spite of these activities and the international debate about energy efficiency and climate protection, building owners in emerging countries still tend to avoid, for cost reasons, investment in the improvement of existing facades with respect to energy conservation, acoustic insulation, and other factors. The long-term and durable value of a flexibly usable property—both material and intangible—provides a strong argument for building owners and investors to support the revitalization of buildings that are major elements in the cityscape. In order to lead this discussion successfully on an international level, the specific character and value of these Modernist buildings must be pointed out. An improvement of the energy parameters and/or internal comfort levels would provide buildings such as those by Oscar Niemeyer in Belo Horizonte and São Paulo with a renewed and long-term lease of life.[XII]

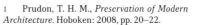

I   Prudon, T. H. M., *Preservation of Modern Architecture.* Hoboken: 2008, pp. 20–22.

II   Blum, H.-J. et al., *Doppelfassaden.* Berlin, 2001.

III   Pottgiesser, U., *Fassadenschichtungen – Glas. Mehrschalige Glaskonstruktionen: Typologie, Energie, Konstruktionen, Projektbeispiele.* Berlin, 2004.

IV   Voelckers O., *Bauen mit Glas. Glas als Werkstoff, Glasarten und Glassorten, Glas in Bautechnik und Baukunst.* Stuttgart, 1934.

V   Oswalt, P. (ed.), *Wohltemperierte Architektur. Neue Techniken des energiesparenden Bauens.* Heidelberg, 1995.

VI   Dorsemagen, D., *Büro- und Geschäftshausfassaden der 50er Jahre. Konservatorische Probleme am Beispiel West-Berlin.* Technical University Berlin, Architecture: Dissertation, 2004.

VII   Rauterberg, H., "Wie ich versuchte die 60er Jahre zu lieben". In: *DIE ZEIT* No. 11, March 11 2010, p. 47.

VIII   Ebbert, T., *Re-Face. Refurbishment Strategies for the Technical Improvement of Office Façades.* Technical University Delft, Architecture: Dissertation, 2010.

IX   Russig, V., *Gebäudebestand in Westeuropa.* IFO Schnelldienst, vol. 52, issue 12, 1999, pp. 13–19.

X   Pilkington K Glass™ is the brand name of the Pilkington Group Limited.

XI   *Die Neuen Architekturführer No. 100.* Berolinahaus Alexanderplatz Berlin. Berlin, first edition 2007.

XII   Pottgiesser, U., "Revitalisation Strategies for Modern Glass Facades of the 20th century." In: *Proceedings STREMAH 2009.* Eleventh International Conference on Structural Studies, Repairs and Maintenance of Heritage Architecture. Southampton, 2009.

14   Europa Center in Berlin: The subconstruction and internal sun shading of the double facade is visible during the renovation in 2002.

15,16 The former BEMGE office building by Oscar Niemeyer in Belo Horizonte: To counteract internal overheating, the building has been retrofitted with air conditioning units—which has significantly affected its appearance.

CONVERSION

A building is adapted for a new use because the old one has become obsolete. The new perspectives for use resulting from the conversion are often the prerequisite for the survival of an old building.

Converting buildings with moderate effort is generally more economical—and almost always more environmentally sustainable—than new construction. Adaptive reuse does not, however, only mean successfully putting new uses into an old shell.

At best, the impression is given that a building, at the moment of its conversion, has finally achieved its true destiny: an almost six-hundred-year-old outbuilding at Waldsassen Abbey, which has stood unnoticed next to the monastery entrance for years, becomes—once completely repaired and freed of disfiguring additions—a stylish cultural center with hotel and restaurant. In Tallinn, the ruins of a rugged factory building prove to be the ideal basis for a stately residential and office tower, a symbol of new economic prosperity. In all the examples featured here, the traces of the past become aesthetic backdrops for newly established uses. Ideas for adaptive reuse uncover the potential of buildings and give them second lives.

FAHLE BUILDING
— KOKO Arhitektid

1|2

3|4

1   From paper factory to residential and office tower: the Fahle Building in Tallinn

2,3   In the peripheral area characterized by old industrial relics, the converted factory is a symbol of a new start—it stands prominently along the road to the airport.

4   The disused paper factory before the renovation

GLASS HEAD ON A STONE TORSO

CONVERSION OF A FORMER PAPER FACTORY INTO A RESIDENTIAL AND OFFICE TOWER
KOKO ARHITEKTID, TALLINN

Like a stone watchman, the old cellulose and paper factory stands conspicuously on a hill above the Estonian capital, Tallinn, directly on the arterial road to Tartu and the airport. The building, named after Emil Fahle, the one-time factory director, dates from 1926 and was in operation until the beginning of the 1990s. Then it was left vacant, after which the tower building with limestone walls of up to 1.2 meters in thickness deteriorated visibly. In the winter of 2001, a group of Estonian architects met in the old factory for to brainstorm and work out possible ways to use the impressive industrial building. They came up with the idea of creating a cultural factory as an event location, with galleries, studios for artists and musicians, and a new venue for the Tallinn theater. In fact, the Estonian Academy of Arts expressed interest in setting up a branch in the building. Triin Ojari, editor-in-chief of the Estonian architecture magazine *Maja,* recalls: "The stairs to the upper floor had no railings any more and it was unbelievably cold in the high void—but we were all fascinated by the mysterious charm and the endless corridors of this giant industrial space."

The vision of a cultural factory was not destined to become reality. At least the base of the building now accommodates a renowned art gallery, restaurants, and a fitness studio, while one of the largest tenants on the office floors is a television studio. The majority of the rest of the building, however—almost two-thirds of the total area—is occupied by apartments; those on the west side have a fantastic view over Tallinn and the Baltic Sea. Thus the building, converted according to a design by KOKO architects, brings a new emphasis to this hitherto dismal peripheral area of Tallinn. KOKO's architectural concept is convincing in its clarity: By adding an additional six stories on top of the original building's eight stories, the prominent location and the tower-like character of the imposing building have been emphasized and—in two senses of the word—overlaid.

The transition to the newly superimposed stories at the ninth floor is marked by setting them back one meter behind the edge of the stone facade. The setback elegantly establishes a spatial joint between old and new, and also corresponds to the two main cornices of the historical building that serves as a base.

For the double-skin facade of the tenth to fourteenth floors, glass in three slightly different tones of green has been used. As a result, the windows break up the light, depending on the incidence of the light and one's viewpoint, into various colors, and the large rectangular glass box shimmers arcanely in the sun—an intentional contrast to the distinct, slightly crude appearance of the rough stone base. Like pixels of color, the differently-tinted glass panes endow the inherently smooth façade with vitality and freshness. One 110 apartments, each 30-145 sq m in size, have been created in the building, of which two-thirds are in the glass superstructure. Depending on the fit-out and views, they cost up to 3,500 euros per square meter.

The new floors, forming a glass head upon a stone torso, transform the building into a landmark visible from afar. The new floor levels are thereby undeniably recognizable as a contemporary addition. The refurbishment of the dilapidated factory site could not have been executed more confidently or comprehensively. All of a sudden, loft living is also en vogue among the young and successful in Tallinn—of whom there are quite a few. Nearly all the apartments were sold at lightning speed.

The conversion of his factory into a business center and panoramic loft building would have pleased its builder, Emil Fahle. He was a genuinely self-made man: In 1895, the son of a railroad engineer had immigrated from Germany with just five rubles in his pocket and had begun working as a factory laborer. Fewer than five years later, at only twenty-four years of age, he had risen to become director of the same factory. FPJ

CROSS SECTION SHOWING THE BUILDING WITH ADDED STORIES

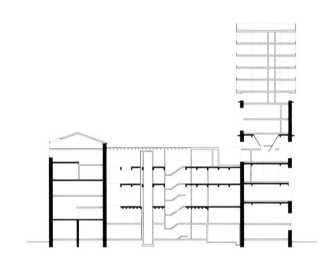

PLAN OF THE 6TH FLOOR

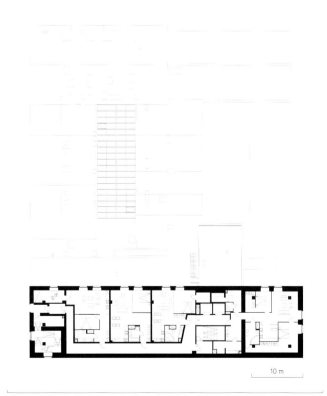

10 m

PROJECT DATA
Client Koger Kinnisvara
Construction costs EUR 15 M
Building footprint 2,568 m²
Usable floor area 16,100 m²
Use 60% apartments, 25% offices, 10% commercial, 5% parking
GFA 19,400 m²
Completion 2007
Project management Raivo Kotov, Andrus Kõresaar
Interior design Liis Lindvere, Raili Paling, Liisi Murula (KOKO Arhitektid)
Location Tartu Road 84a, EST–10112 Tallinn

Year of construction 1926

1000   2010

Conversion 2007

2000   2010

Construction costs   **Price per m²**
15 M €   932 €

15,000

10,000

5,000

0

5   Office unit with conference room
6   Piranesi's paper factory: the
interior before starting the rehabilitation work

7   One of the circulation areas
8,9   The industrial history of the building still has a presence in many apartments: silo hopper on the 5th floor

ANDEL'S HOTEL LODZ
— OP Architekten

1   The view along the facade illustrates the enormous dimensions of the seven-story, 200 m long factory building. The hotel's swimming pool cantilevers beyond the facade as a glass box.

In 1810, Łódź was a sleepy town of some 200 inhabitants. Less than ninety years later, in 1897, 314,000 people were living in the city, which to this day is the second largest in Poland. Łódź was the boom town of late nineteenth-century Poland—the country's equivalent of Manchester—with all the negative side-effects of capitalism, including several workers' uprisings that were bloodily suppressed. Łódź, however, was also the city in which Poland's first movie theater opened in 1899, and where Poles, Germans, Russians, and Jews lived together in a bustling and natural form of coexistence. The textile industry had transformed Łódź into a metropolis; in 1904, there were 546 factories with 70,000 employees.

POLAND'S MANCHESTER BECOMES CHIC

CONVERSION OF A FORMER TEXTILE FACTORY INTO A HOTEL
OP ARCHITEKTEN, VIENNA

The floor slabs have been cut away along the longitudinal axis of the building to form an elliptical inner court, extending through all seven levels. This court introduces a vertical accent, opening up an architecture that is otherwise defined mainly by horizontals.

The second major intervention has been made on the outer shell and is the eye-catcher of the restored building: A transparent box boldly cantilevers out from the roof terrace over the main facade. Inside it is the swimming pool belonging to the wellness area.

PROJECT DATA

Client Warimpex Finanz- und Beteiligungs AG
Construction Costs EUR 70 M
Usable floor area, total 33,300 m²
GFA 40,100 m²
Number of hotel rooms 220 rooms and 58 apartments
Feature 2009 European Hotel Design Award 2010 *Contract Magazine* Interior Award in the category "Adaptive Re-Use"; awards from the real estate industry incl. a "Special Tribute to the Guest of Honour" at the 2010 MIPIM Awards.
Completion 2009
Architecture/project management OP ARCHITEKTEN, Wojciech Popławski, Andrzej Orlinski
Interior design Jestico + Whiles
Location Ul. Ogrodowa, PL–91065 Łódź

Year of construction 1887

1000 ———————————————— 2010

Conversion 2009

2000 ———————————————— 2010

Construction costs   **Price per m²**
70 M €   2,102 €

15,000
10,000
5,000
0

One of the largest of these was the factory belonging to Izrael Poznański. In 1887, the entrepreneur founded a weaving mill that is reminiscent of an ocean liner in size—whereby the brick facade is not smooth, but entirely articulated by pilasters, masonry corbels, and cornices. It was an industrial hall full of noisy machines, but with the facade of a monumental building. Almost 200 m long and 33 m high, the colossus of red brick had seven floors with a total of 40,000 sq m of production area.

The factory continued to be used for industrial purposes until into the 1990s—then it stood empty. Demolition was certainly the most likely prospect for the ensemble—after all, how does one convert a building of these dimensions?

Today, the building meticulously restored still bears testimony to the tumultuous boom phase in Łódź, thanks to the audacity of an Austrian investor and the enthusiasm of Wojciech Popławski and Andrzej Orlinski of OP Architekten. They have demonstrated great sensitivity in making a reality of the investor's proposal to transform the massive brick building into a four-star hotel with all the requisite amenities.

The sheer amount of space proved not to be an obstacle to this goal, but rather a trump card for the revitalization. There was not only enough space in the building for 278 rooms and suites, but also for a 3,100 sq m conference center on the ground floor, a spa with swimming pool and wellness landscape on the top floor, and a ballroom with space for 800 guests. At 1,300 square meters, this takes up about half of the fourth floor, and is acoustically separated from the rest of the building. The 7 m-high banquet hall is one of the largest in central Poland. Rounding off the facilities are bars and a restaurant, located on the ground floor and the top floor. Diners in the restaurant find themselves in an expansive, but not over-dimensioned hall, amidst 4 m-high cast-iron columns; no suspended ceiling obstructs their view of the shallow Prussian vaulting—a cast-iron, steel, and brick construction typical of buildings from the late nineteenth century. Wherever possible, for instance in the stairways and in the lobby area, the architects have retained and restored the bare shell of the industrial architecture and have subtly embellished it with contemporary elements. In so doing, they have created an intriguing contrast to the dignified interiors produced by the London designers Jestico + Whiles for the guest rooms and part of the dining area.

CROSS SECTION THROUGH THE BUILDING AFTER RENOVATION

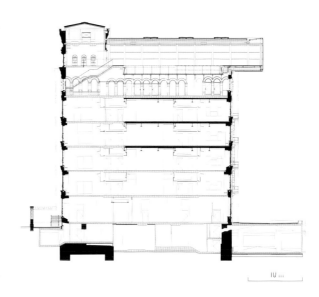

10 m

SECTION THROUGH THE GROUND FLOOR LEVEL

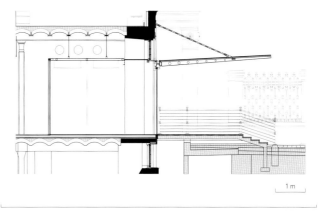

1 m

The swimming pool was integrated by the architects into the top floor of the building within a void formed by a cast-iron water tank. The walls of the old tank are discernable through the glass floor, to the left and right of the pool. It was built 130 years ago in Manchester and integrated into the roof of the factory as part of a fire extinguishing system that was advanced in its day. When they look out from the pool projecting freely into the sky, swimmers enjoy a panoramic view across the roofs of the city. At a dizzying height, interior space and urban landscape merge—the water in the pool and the skyline of Łódź unite. Those who enter the upper floor of the old factory, today, come here to enjoy the city; the city of the workers is history. FPJ

2

3

2   Oval slab openings in the center of the building connect the floor levels.
3   Hotel lobby beneath Prussian vaulted-brick ceiling: the industrial building reveals unimagined elegance.
4   The event hall with conference seating
5   Living room in one of the suites
6   Spa and swimming pool on the roof of the hotel
7   The roof terrace

STRUCTURAL CHECK FOR THE BUILDING'S SKELETON
— Working on Existing Buildings—Experience Gained by a Structural Engineer

A design manual should give its readers help and inspiration on how to tackle certain tasks and solve certain problems. In structural design, a panacea approach should be warned against. In my twenty-five years of practice I have often come across identical-looking defects which, as it turned out, had different causes. However, sustainable refurbishment can only be achieved by addressing the causes of any defects. For this reason all work must start with a case history followed by an analysis of the strengths and weaknesses, so that one works with the existing structure, rather than against it.

RAINER HEMPEL

CASE HISTORY
—

It is generally of advantage to approach a problem systematically rather than to jump to conclusions. We always start with the case history:
All available information about the history of a building and its construction process is compiled, organized, and evaluated.

The more of this background information is available, the more successful and economical will be the refurbishment or other building work. It is very important that the drawings available are checked against the existing buildings, at least in parts. For example, when revitalizing the former Hessische Landesbank in Frankfurt, we were able to secure excellent drawings of the existing building.

They included the complete set of detailed drawings, as submitted for building control approval, as well as formwork drawings. However, when we sample-checked the existing building against the drawings we found that the openings in the reinforced concrete cores did not coincide with those in the approved formwork drawings. The lift door openings had been made larger and the penetrations for building services had also been modified in size and location compared to the drawings—a clear indication that in this project, design proceeded while construction was in progress. It follows that more recent drawings had to exist, or at least had existed—in addition to those available to us—because it would not have been possible to change the formwork for the reinforced concrete cores in a project of this size by verbal instruction on site.

The new geometry of the formwork design and associated modifications to the reinforcement layout needed detailed drawings for their execution. After extensive research and quite persistent enquiries we found the documents we were looking for in the archive of the company which carried out the shell construction. In this way it was possible to save the building owner significant financial expense as well as the considerable delay of a detailed survey and material investigation.

DIAGNOSIS
—

The second step of our approach is the diagnosis. In this step we carry out an evaluation. Here, it is important to distinguish between the cause and effect of defects. Analyses are produced of the strengths and weaknesses in order to produce a "constructive"—in both senses—decision-making aid for the architect/designer. This part is the most creative phase for the structural engineer.

THERAPY
—

The third step we refer to as "therapy" and includes the documentation of structural integrity. With regard to historical buildings and those with serious damage, the therapy step is extremely important because it is the prerequisite for a successful solution, whether this includes refurbishment, revitalization, or shoring-up the building.

SAMPLE PROJECTS
—

It is the task of the structural engineer to demonstrate the structural integrity and fitness for purpose of buildings by means of structural calculations. For new buildings we have at our disposal a comprehensive body of regulations, standards, structural theories, and measuring methods for building components, connecting methods, materials, etc. The regulatory instruments applicable today have only limited relevance, if any, for older buildings. Builders in the past had their own rules and regulations which they followed more or less thoroughly. When assessing a building, we therefore can not avoid taking into account the standards and regulations in force at the time of the construction. Naturally, we also need to consider the intended use of the building, particularly the resulting imposed loads.

A current and typical example of this scenario is one of our projects, the Stockwerksfabrik in Bad Hersfeld. Essentially, this consists of an early ferroconcrete construction built in 1910. This new material, ferroconcrete, was used with very economic, slender sections: The floor slabs were designed for imposed loads of 800–1000 kg/m^2 (today 8–10 kN/m^2). Unfortunately documentation for the structural design, such as structural calculations, position drawings, and formwork and reinforcement drawings are no longer available. This meant that we had to carry out a thorough investigation and case study, involving the recording of defects, investigation, sample-taking, evaluation, etc. of materials. We discovered what, by today's standards, are considered serious defects: For example, hoops in concrete beams were generally not closed. Again, according to current standards, the existing deflection and shear reinforcement was clearly inadequate in certain places, which meant that numerous reinforcements would be necessary.

In this context it is necessary for us to take a critical look at our current professional thinking. If we think "How did our colleagues in 1910 manage to do something so 'inadequate,'" we fall prey to a subjective view that is totally rooted in our present-day thinking.

Conversely, our colleagues from 1910 would probably be quite surprised by our presumptuous criticism.

An excursion into building history and particularly, for structural engineers, into the history of building construction, helps to gain a perspective on this seemingly unbridgeable difference of views: Reinforced steel concrete construction is still a relatively young construction method. The first ferroconcrete designs appeared in Germany shortly before 1890. However, the first regulations and measuring methods were not established in Prussia until 1905. Therefore the application of ferroconcrete, which at the time was still a new material, was still in its infancy and that applies to the manufacturing processes as well as to the construction rules and measuring methods.

Theoretical and practical recommendations for constructions with this material that reflect the latest state of technology around 1900 are linked to the names Koenen and Mörsch. Above all, the standard work by Emil Mörsch *Der Eisenbeton – seine Theorie und Anwendung*[I] provided contemporary designers with an up-to-date technological basis.

When we look at these regulations from around 1910 we find that closed hoops were not yet included. The designs considered inadequate by us were in compliance with scientific standards of 1910 and therefore represented best practice.[II]

A loadbearing construction which has served its purpose for a hundred years without detectable defects has proved in practice that the associated regulatory framework was appropriate. Of course, today we have the facility of using the finite-element method on the computer for visualizing and determining all structures spatially and more realistically. And yet we are still not in a position to determine the structural integrity of designs completely without any doubt.

So how can we find a satisfactory solution to the elementary building code requirement for documenting the structural integrity of buildings?

In practice, we employ several procedures: The most elegant option is the method involving what is called "relative safety." In this case we start from the premise that the existing structure has fulfilled its loadbearing function with existing imposed loads for several decades. If no detectable defects have occurred for several decades, then clearly the factor for structural integrity is greater than 1. Even if certain defects can be detected the factor for the structural integrity is still at least 1. In such cases it makes sense to work towards increasing structural safety to a defined level, possibly by means that are detached from the original loadbearing structure. One such means would be to reduce the loads imposed by changing the future uses of the building. Another way of reducing the loads and resulting structural stresses would be to remove floor finishes suspended ceilings and suchlike. In this way it is often possible to increase

RAINER HEMPEL
—

Prof. Dr.-Ing. Rainer Hempel, born in Neustadt a. d. Orla (Thuringia), studied civil engineering at the TU Braunschweig. From 1979 to 1981, he was a structural engineer in the practice of Dr. Rehr und Martin, Braunschweig. From 1981 to 1987, he was a research assistant in the department of structural analysis at the TU Braunschweig. In 1981, he founded the engineering firm Hempel & Partner in Braunschweig. He gained his doctorate (Dr.-Ing.) in 1986. From 1989 to 1991, he was a professor at the University of Siegen. Since 1991, he has been a professor at the Faculty of Architecture, Cologne University of Applied Sciences; he was dean of the faculty from 1998 to 2002. In 2004, the practice Prof. Dr.-Ing. Hempel & Partner was relocated to Cologne. The practice offers services across the entire range of structural engineering, in particular it develops innovative and creative structural concepts for revitalizing existing constructions and rehabilitating historical load-bearing structures, as well as for new buildings.
www.hempel-ingenieure.de

the existing structural safety factor by up to ten percent. By inserting additional loadbearing members it is possible to determine the structural safety relatively easily and accurately. We should not forget, however, that a design and construction does not "know" how we have calculated its dimensions. It will always react to the forces and physical laws at work in its actual construction. This means that additional structural members will only take part in the transfer of loads if their degree of deformation is identical with that of the existing members. For this reason, reinforcements should usually be installed pre-tensioned. In such a case they will contribute immediately to the loadbearing function of the existing structure and, through their additional deflection, will directly aid the transfer of imposed loads.

Another method for assessing the loadbearing capability of a structure is to impose loads in a trial until breaking point is reached. This is an excellent method where loadbearing members can be tested up to breaking point. Such trials can not be carried out on site but require the facilities of a test laboratory. In order to arrive at statistically reliable results it is necessary to run a number of tests. Such a test method can only be used where sufficient structural elements are available which are no longer needed in the future, for example in cases where additional staircases or lifts are installed.[III] Trial tests with loads that are clearly below breaking loads can also be carried out on site. It must be said, however, that in this case very sophisticated deflection measurements are required. With steel-reinforced concrete and ferroconcrete constructions it is important to ensure that the steel is only stressed in its elastic range and does not undergo plastic deformation.

REVITALIZATION OF AN OFFICE AND RETAIL BUILDING IN BERLINER ALLEE, DÜSSELDORF [IV]
—

This building is a concrete frame construction dating from 1954. Complete drawings from the time of

construction were available. The structural dimensions complied with regulations in force at the time so that the building owner did not see the need for an intervention. However, it turned out that there was no documentary evidence of the structural safety of the building with regard to bracing. During our case study and our first visit on site we noticed that the concrete surface was very powdery, with fine dust sticking to our hands when we touched it—an indication of clay components in the concrete aggregate. These inhibit or prevent the grain formation that concrete needs to develop its full strength. As a result of this suspicion that the concrete was of inadequate quality, we decided to take core samples. Tests of the samples showed that instead of a concrete quality of B 225 or B 300, the existing concrete was only of B 80 quality, which corresponds to C 8 in the more recent classification method, a concrete

quality which even at the time of construction was not permitted for steel-reinforced concrete. Such results are truly surprising. However, we know that up until the 1960s, it was not uncommon for aggregate materials to be directly obtained from gravel pits. It was therefore not uncommon to find various grades of clay in the aggregates. It was not until about 1960 that aggregate materials were sieved and washed so that the materials could be graded to defined grain diameters and hence used to produce high-quality concrete.

Owing to the low compressive strength of the concrete it seemed obvious to use a loadbearing imposed concrete layer to act in combination with the existing cross section. The bending pressure zone was increased by five centimeters by adding a C 20/25 grade concrete (concrete with a greater difference to the existing C 8 grade would have been detrimental), which resulted in a significant improvement in loadbearing capacity. The practicality and effectiveness of such a method in terms of function and practicality needs to be ascertained and

NEW BONDED CROSS SECTION OF THE UPGRADED FLOOR SLABS WITH CONCRETE TOPPING
A loadbearing concrete layer on top reinforces the existing construction of the Berliner Allee office building.

Section A–A with support reinforcement

1 m

Section B–B in a typical field

1 m

A Ø 10 - 9
B Concrete topping C20 / 25, h = 5 cm
C Q188A laid directly on the existing slab, but with bearing bars oriented parallel to the lettered axes

D Ø 10 - 50, 4 per m², drilled-in with Hilti Hit HY-150 or equal, installation in accordance with approvals
E Existing slab

1   [p. 144] **Viktoria Versicherung AG commissioned the revitalization of this office building dating from 1954. Condition prior to refurbishment**

2   [p. 144] **View of the remodelled facade following refurbishment by the architects Bartels and Graffenberger**

documented. For this purpose we asked for test areas in the building to be prepared as follows:
- Sand-blasting the concrete surface/exposing the grain structure
- Laying the reinforcement directly on the existing floor slab
- Fixing plugs mechanically with the drill and glue method including stirrup caps that completely enclose the additional reinforcement
- Removing concrete dust, loose stone, and aggregate by vacuum suction
- Applying an epoxy resin bonding layer by spray application
- Applying the bonded concrete with plasticizer and grain diameter limitation
- Applying this concrete while the bonding layer is still fresh
- Treating the concrete surface to meet tolerance levels for smoothness
- Finishing the concrete surface

In order to support the load of the freshly-applied concrete and reduce the deflection resulting from it, the floor slabs had to be supported with temporary props through three floors.

BENNO SCHILDE PARK, BAD HERSFELD[V]
—

After the Babcock-BSH machine factory (formerly the Benno Schilde company) moved out of its premises to the north of the old town centre of Bad Hersfeld in 2009, the town council was presented with a unique opportunity to acquire the property and establish a recreational space on the former industrial land.

The production buildings and the office block were removed, but the Stockwerksfabrik and former assembly building to the west, which both enjoy landmark preservation status, were retained (see site plan).

The re-zoning of the grounds to form a city park with cultural and educational functions at its center was further enhanced by the renaturation of the small River Geis. When the factory had been constructed, the Geis had disappeared underneath a reinforced concrete slab.

CONVERSION OF THE DISUSED STOCKWERKSFABRIK TO A SCIENCE CENTER AND EDUCATIONAL ESTABLISHMENT
—

Until the end of 2009 the factory was used as a storage building. It consists of a northern wing without a cellar and a southern wing with a cellar. Both wings have three full floors. The different parts of the building were constructed at different times and differ in their design and materials.

The northern wing, built in 1904, is a mixed construction: the richly embellished brick facade conceals a multi-story timber structure. All floor decks are of timber beam construction supported by two timber beams running along the length of the building. These in turn are supported by timber posts at regular intervals.

GROUND FLOOR PLAN OF THE BENNO SCHILDE FACTORY AFTER CONVERSION

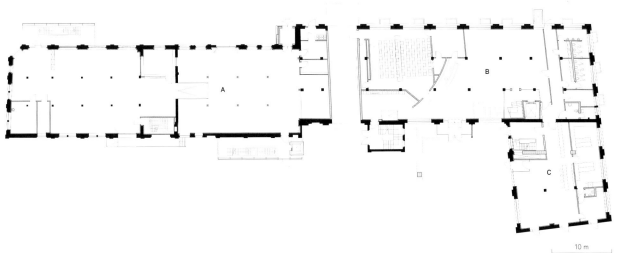

A Teaching workshop
B Science center
C Restaurant

Between 1910 and 1913, the southern wing of the Stockwerksfabrik was added: a solid masonry construction, likewise featuring richly ornamented brickwork facades with pilaster-type elements. The inner load-bearing structure was erected using the most modern building material available at the time—ferroconcrete. As with the northern wing, two supporting beams run the length of the building, once more creating a three-bay layout. At right angles to the main beams are secondary beams which support the slab with thicknesses of 10 and 12 cm. The main beams are connected to the ferroconcrete pillars via coved connectors.

In view of the required uses, the southern wing had been designed for relatively high imposed loads of 800–1000 kg/m² (today 8–10 kN/m²). The author's case study did not reveal any serious defects. However, it was clear that the reinforcement did not have sufficient concrete coverage as it was exposed in some places. In order to comply with our duty of care, we had material samples taken and opened up some parts of the construction.

The results were surprising. With a quality of B 225 the concrete quality was what we had expected, comparable to our current C 20/25. The amount of reinforcement corresponded approximately to that required under current standards—except for the shear and hoop reinforcement. Emil Mörsch, the leading ferroconcrete researcher at the time, made the following statement: "As was demonstrated in the trials described later on, the hoops only have secondary significance for the shear strength of the beam so that their use is determined by the practical considerations of a secure connection of the beam with the slab and ensuring their friction resistance."[VI]

This was the principle the designers had followed. Other design rules at the time contained provisions which had the effect that certain reinforcements were not placed where we would expect them to be fitted under today's rules. For example the coved connectors with the main beams were not reinforced and therefore, in principle, do not make a contribution to the load transfer. However, as they are located on the pressure side of the beams' deflection zone (below the beam, at the pillars), they can be taken into account to a certain degree even though the pressure zone is not enclosed by hoops. There is also the fact that they did not show any cracking at all.[VII]

With the help of numerous comparative calculations taking into account the local conditions and the modification of various structural models, we were able to provide evidence of the structural safety of the building, albeit quite a way off current standards. In such cases it is essential that the authorities in charge of supervising the Building Code have an open attitude towards the structural engineer's reasoning and are prepared to go along with it.

CROSS SECTION THROUGH THE PRODUCTION HALL AFTER CONVERSION
After individual reinforcement measures it was possible to retain most of the slender structure of the original design.

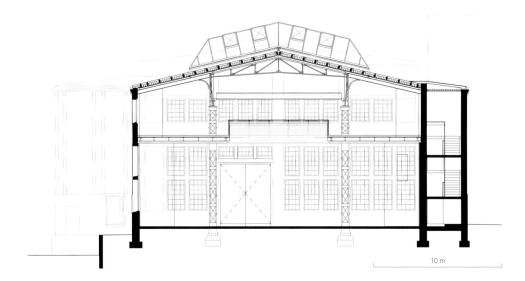

10 m

3 The upper floor of the factory, showing its state when industrial use ceased

4 [p. 144] The former production hall was the central focus of the whole complex.

CONVERSION OF A FORMER
ASSEMBLY BUILDING
TO AN EVENTS VENUE

—

The assembly building of Works II of the Benno Schilde machine factory is a typical three-bay hall construction, which was built in 1912. The solid external walls are built in brick and line the inner steel frame con-

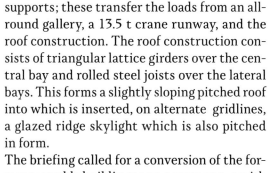

struction with its two rows of lattice girder supports; these transfer the loads from an all-round gallery, a 13.5 t crane runway, and the roof construction. The roof construction consists of triangular lattice girders over the central bay and rolled steel joists over the lateral bays. This forms a slightly sloping pitched roof into which is inserted, on alternate gridlines, a glazed ridge skylight which is also pitched in form.

The briefing called for a conversion of the former assembly building to an events venue with a fixed stage and all modern visitor service facilities. The link building houses the artists' dressing rooms, as well as other functions.

The conservation plan for the former assembly building aimed for a careful overhaul which would retain traces of its former use on wall and pillar surfaces and preserve the rough character of the industrial building as backdrop to the new functions.

During our case study we discovered that there is no functioning bracing system in place for the building. In consultation with the architects we initiated an upgrade installation of crossing diagonal bracing members in the roof and at gallery level. These have the effect of safely transferring wind and other lateral loads into the ground via vertical walls on gridline 2 and vertical members on gridline 9.

Further serious defects and damage were evident in the skylights and the timber construction of the roof skin, which meant that both these components had to be renewed. The new use of the building meant that the roof construction had to satisfy more stringent requirements than before. The slender lattice girders were not adequate for the additional

loads from new insulation, double-glazing, and the new, heavier construction of the skylights. However, our objective was to retain as much as possible of the historical substance and original character of the building. In order to come as close as possible to meeting this objective, our office, in cooperation with Kleineberg and Pohl, the architects responsible for the conversion concept at Bad Hersfeld, developed a special principle which we call two-layer construction: This model is based on the assumption that the existing structure performs part of the load transfer in accordance with its existing structural properties and design and that a second structural layer is introduced for the additional loading, including any excess. This additional structure is also shown as such, either by selecting cross sections, materials or profiles which were not available at the time of construction, or by a specific color scheme.

In this case the structural engineer has supported the load from the new skylights on steel profiles, which transfer the load to the existing supports of the crane runway. This load thus no longer needs to be borne by the slender triangular lattice girders, allowing them to continue performing the function for which they were designed.

I   Verlag Konrad Wittwer, Stuttgart 1906.
II   The same applies to the field reinforcement of slabs and beams which, according to today's regulations, has to be extended by at least one third beyond the front face of the support. In 1902, it was still common practice in single field slabs to fold up the lower reinforcement layer at approx. 0.2 × l in front of the supports and continue this as upper reinforcement above the support, which causes the element to be restrained. In this way the calculated field moment was reduced from ql2/8 to one third, i.e. ql2/24. The resulting support moment was therefore ql2/12.
III   At the locomotive hall project in Göttingen we were able, in spite of the low compressive strength of the pumice concrete slabs from 1920, to determine a safety factor of approx. 2 compared to the breaking load, for the load case of own load + new insulation + sealing profile + one man load in the centre of the field or, alternatively, exposure to full snow loading.
IV   Building owner: Victoria Versicherung AG, represented by MEAG, Munich / Architects: Bartels and Graffenberger, Düsseldorf.
V   Building owner: Wirtschaftsbetriebe Bad Hersfeld GmbH, Bad Hersfeld / Architect: Kleineberg and Pohl, Braunschweig / Landscape architects: Wette + Küneke, Göttingen.
VI   From: *Der Eisenbeton—seine Theorie und Anwendung* (Ferroconcrete—Its Theory and Practice), Stuttgart, 1906, p. 9.
VII   In addition there was the fact that the continuation of the main beams was only partially effective because the reinforcement of the uprights is very short and consists largely of continued folded-up shear reinforcement.

SUMMARY

—

In my view, the successful and economical execution of revitalizations, conversions, and extensions is dependent on some essential prerequisites:

- Careful case study and diagnosis together with an analysis of strengths/weaknesses prior to the design process
- A design team with an integrated approach to the work
- A creative approach to the existing building, particularly its loadbearing structure Crucial prerequisites are an understanding of the history of construction and relevant experience with historical and current building materials
- A critical appraisal of standards, rules, and building codes
- The readiness to devise and defend innovative solutions
- Building owners who are willing to support your approach with their decisions.

5,6   [p. 144] In the future it is intended to use the former production hall as an auditorium: model of the converted factory

GYMNASIUM IN A HANGAR
— Numrich Albrecht
Klumpp Architekten

1   View from the grandstand to the hall
2, 5 Long span: The coffered,
reinforced-concrete structure of the
roof is an eye-catcher.
3   Corridor leading to the locker rooms
4   Foyer and the gym attendant's office

AVIATION HISTORY MEETS AMATEUR SPORTS

GYMNASIUM IN AN OLD AIRPORT HANGAR IN BERLIN-ADLERSHOF
NUMRICH ALBRECHT KLUMPP ARCHITEKTEN, BERLIN

It's a Wednesday morning in the south of Berlin: Ten vigorous seniors are standing in a semicircle around their trainer in the middle of a gymnasium, tapping red balloons lightly to keep them in the air. The long hangar is somewhat older than the seniors on the court. Eighty years ago, in the pioneer days of aviation, airplanes parked beneath the wide span of the concrete envelope—which explains the extraordinarily spacious interior, even for a gymnasium. The hangar is a relic of Johannisthal airfield, which was abandoned back in 1952. Today, the hall with reddish brick facade is surrounded by institutes of Humboldt-Universität, which uses it jointly with the borough.

The architects Numrich Albrecht Klumpp have converted the hangar into a four-court gymnasium. Although it was built for an entirely different purpose, the building offers space for two full-size courts or four smaller courts running crosswise for training purposes.

The conversion of the old hangar has proven to be a stroke of luck for landmark preservation: After numerous interim uses, the borough administration had considered demolishing the preservation-worthy building. But since a gymnasium was already planned for the neighborhood, the architects were able to convince the building authority that the hangar could easily be transformed into one.

Not only are the dimensions of the interior space impressive, but the structure also has a certain beauty: the gently vaulted coffers of the prestressed concrete ceiling slab as well as the powerful longitudinal girder, together with its bracing, that runs along the bleacher side and was revealed during the conversion. Below this girder, rolling shutters once extended along the entire length. The playfulness and mastery with which the architects unite old and new is impressive—they only intervened in the built substance where it was useful and necessary: They removed the remains of the shuttered gates, exposed the reinforced concrete truss above, and inserted a new tract into the resulting opening—for lockers, toilets, gymnastics rooms, and a foyer. The main facade is glazed continuously along the ground floor, and the first floor is characterized by a continuous window band. The adjoining new building opens to the hall with a spectators' gallery and an open access corridor at the upper level. The main entrance is located in the middle of the new building's main facade. From the foyer, a stair leads to the upper floor. Here, visitors may be surprised to find themselves not only gazing into an immense hall, but standing again in a daylit foyer—adjoining the staircase that leads to the gallery and gymnastics rooms is an inviting roof terrace. Just like the 199 seats of the spectator bleachers, the terrace was not initially required by the clients, yet the architects convinced them that such deviations from the program would augment the hall's comfort and its range of use. Still, the budget remained within the established limits of 3.9 million euros.

PROJECT DATA

Client Bezirksamt Treptow-Köpenick, Berlin
Construction costs EUR 3.9 M (cost groups 200–700)
Usable floor area 2,920 m²
Gross volume 32,300 m³
Feature Retention of the historical load-bearing structures
Completion 4/2008
Project management/team Arthur Numrich, Jessica Voss
Location Merlitzstraße 16, D–12414 Berlin

Year of construction 1929

1000 ———————————————— 2010

Conversion 2008

2000 ———————————————— 2010

Construction costs   **Price per m²**
3.9 M €   1,336 €/m²

15,000
10,000
5,000
0

GROUND FLOOR PLAN

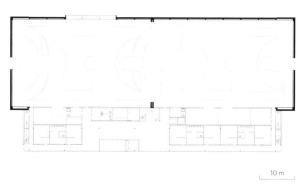

10 m

UPPER FLOOR PLAN

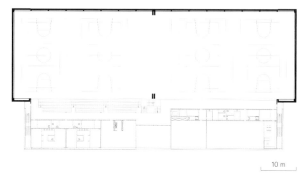

10 m

LONGITUDINAL SECTION THROUGH THE HALL

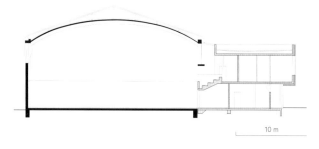

10 m

The new building is so naturally connected with the industrial character of the expansive hall, that it seems never to have been otherwise. The meticulous, simplified detailing is convincing: The radiators are recessed into small niches such that their outer edges lie flush with the plane of the wall. The bleacher seats are polished concrete bases terminated along the top with dark-stained wood. No praise of austerity, but sensible reduction. Whereas sober colors dominate in the hall, with its white ceiling and warm gray shock-absorbing wall panels, strong color accents are used in the new building: The office of the gym's attendant radiates a deep red, and elsewhere there are green and violet walls. The envelope of the addition is a warm anthracite gray. FPJ

Until it was refurbished, the hall was used by a carpet dealer.

6,7 The circulation and service block appended to the side: lockers and washrooms are on the ground floor, gymnastics rooms are on the upper level.

8 View of the hall from the street

9 Entrance, with visitors' terrace above right

WILDAU LABORATORY BUILDING

WILDAU (DE)
OLD Industrial —**pp. 18, 132, 136, 148, 161**
NEW Educational —**pp. 58, 95, 98, 108**

— Anderhalten Architekten

THE LEARNING MACHINE

LECTURE HALL AND LABORATORY BUILDING, HALL 14, WILDAU
ANDERHALTEN ARCHITEKTEN, BERLIN

Hall 14 was a typical mid-nineteenth century production hall on the premises of the former Schwartzkopff factory in Wildau, south of Berlin. Along with a number of new buildings for the Technische Hochschule Wildau, an educational institution newly founded in October 1991, two old assembly halls on the campus have been renovated for academic use: Hall 10, converted previously, accommodates the cafeteria and library, whereas Hall 14 contains the lecture hall and lab rooms.

Roller bearings for locomotives were produced until the 1980s in this well-lit hall with a footprint of more than 4,000 m². The architects have positioned the large lecture hall, seminar rooms, and laboratories—which have specialized technical equipment—within the empty shell in such a way that they are independent of the historical structure. In order for the new use not to impair the unity of space in the factory building, the envelopes of the elements inserted have been made largely transparent. This creates an exciting contrast between the highly technical research laboratories and the neo-Gothic masonry shell.

The original parts of the building have been carefully refurbished. The intention in doing so was not to restore the building to a pristine state, but to leave evidence of different phases of use and past modifications. For instance, built changes to the facade openings have not been reversed. Even after renovation, they remain visible, along with the newly constructed openings, as interventions in the original facade. The brick surfaces have been cleaned on both the exterior and the interior—but the patina testifying to their age remains. A clear symbol of the change in use is the entrance structure placed in front of the main gable on the campus side. The full-height glass-and-concrete form marks the main entrance and opens up the view to the lecture hall behind.

Owing to differing ridge heights in the two parts of the building, the architects have constructed the built-in elements with two and three stories, respectively. As the largest space within the converted building, the lecture hall—which extends over three floor levels and has three hundred seats—determines the internal organization. The load-bearing structure of the new elements is of reinforced concrete; the wall and ceiling panels are of exposed concrete. The new inner facades, consisting of prefabricated, flush-mounted steel-and-glass elements, recall the original industrial use of the site and, in their precision, form a striking contrast to the roughness of the surrounding masonry.

The thermal envelope is formed by the historical facade together with the new inner facade. This creates a thermal buffer zone between the two shells, fed with heat generated from the inner space. The roofing and glazing of the original shell have been insulated. Missing windows in the facade have been replaced by the architects with new, steel-framed, insulating glass windows; the existing windows have single glazing.

PROJECT DATA

Client Federal state of Brandenburg, represented by the Ministry of Finance, Potsdam
Construction costs EUR 16 M
Usable floor area 4,000 m²
GFA 4,250 m²
Feature Building with double shell—the offices, laboratories, and lecture halls inserted into the industrial hall constitute a "building within the building."
Completion 2007
Project management Jürgen Ochernal
Location Bahnhofstraße 1, D–15745 Wildau (Brandenburg)

Year of construction 1902

1000 ——————— 2010

Conversion 2007

2000 ——————— 2010

Construction costs 16 M € **Price per m²** 4,000 €

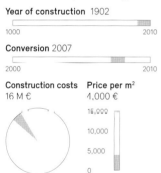

15,000
10,000
5,000
0

CROSS SECTION THROUGH THE HALLS

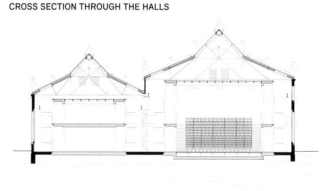

10 m

UPPER FLOOR PLAN

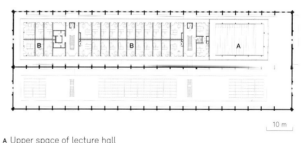

10 m

A Upper space of lecture hall
B Offices (professors/assistants)

Owing to warmer temperatures within the inserted volume, there is no risk of condensation on the elements protruding into the interstitial space. This permitted the floor and ceiling slabs of the galleries to be constructed without thermal breaks.

The renovation of Hall 14 is an exemplary model for the adaptive reuse of a large, historical hall whose space is divided among structures of a smaller scale, while accommmodating a multitude of uses. The landmarked building and its new technical and didactic use complement each other ideally.

Locomotive production in Wildau, historical photo of the factory

Empty hall immediately before renovation

1 [p.152] The main entrance at a
gable end of the larger building
2   Staircase to the upper level
3   A second, inner facade encloses
the offices and workspaces, forming a
thermal enclosure.
4,6 Transparent instruction: the
large lecture hall vis-à-vis the main
entrance
5   Laboratories and offices on the
upper level

3

ALVÉOLE 14

SAINT-NAZAIRE (F)
OLD Bunker
NEW Restaurant/Bar —pp. 28, 92, 136, 168
Cultural facility

CULTURAL CENTER
— LIN Finn Geipel + Giulia Andi

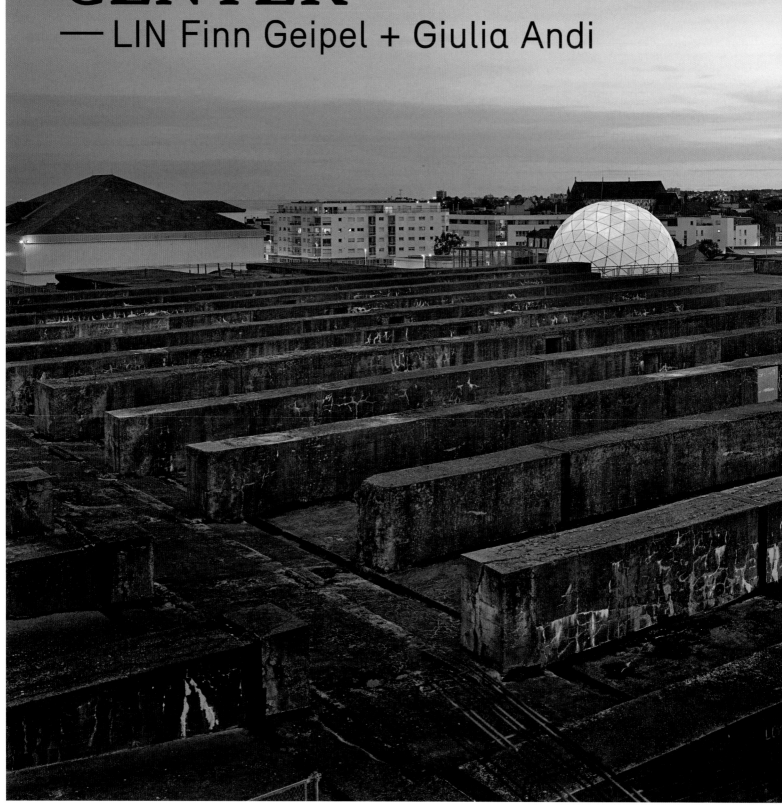

1,3 View across the roof of the former U-boat bunker toward the illuminated dome, which also marks the location of Alvéole 14's roof terrace
2 The dome, made of triangular plastic elements, originally crowned a NATO radar tower (far right) at Berlin's Tempelhof Airport.
4 Saint-Nazaire seen from the harbor: in the middle is the U-boat bunker with its dome.

UP NINETY STEPS THROUGH THE THICKNESS OF THE SLAB

CONVERSION OF A U-BOAT BUNKER INTO A CULTURAL AND MUSIC CENTER, SAINT-NAZAIRE
LIN, FINN GEIPEL + GIULIA ANDI

In 1939, Great Britain declared war on Germany. What gave Winston Churchill, the indomitable British prime minister, particular cause for concern were the German U-boats. They used torpedoes to attack not only the battleships of the Royal Navy, but especially civilian cargo ships, of which they sank hundreds in a brief period.

Their main operational bases were ports in occupied France, in particular Brest, Lorient, and Saint-Nazaire. With growing British defensive action, the defenses of those U-boat bases were continually increased. The U-boat bunker in Saint-Nazaire, constructed from 1941 to 1943 by forced laborers, has a reinforced concrete roof 4 m to 9 m thick. With a length of 295 meters and containing 14 U-boat cells, it is amongst the largest U-boat bunkers in Europe. Whereas the town of Saint-Nazaire was 85% destroyed during the war, the bunker survived largely unscathed—ever since which it has stood as a monstrous and indestructible barrier between the city and its harbor. Aside from small areas that were used for storage and the like, the bunker long appeared to be unusable for upscale functions appropriate to its central location and thus stood empty. In quest of a worthwhile use for the building, the municipality of Saint-Nazaire held an architectural competition in 2003. The first prize was won by Finn Geipel and Giulia Andi for their idea of converting one of the cells into a music and cultural center with event rooms. They called the center Alvéole 14 (literally Cell 14) because it was integrated into the fourteenth cell, at the southern end of the bunker.

Like the other cells, it consists of two parts: a 22 m-long section at the landward end, which was formerly used as a storage area and workshop, and the 92 m long U-boat pen itself, with its basin open to the harbor. Between the two sections is a 5 m-wide service corridor running along the length of the complex, with tracks connecting all the cells together.

The new venue of 5,570 sq m takes up about 9% of the space inside the bunker, and is arranged into two areas: the large, the actual U-boat pen has become a flexible event hall, while the former workshop area serving the 13th and 14th cells accommodates a music venue known as "VIP," along with associated recording studios, a bar, management offices, and technical plant rooms. The bar is located on a balcony above the hall and offers the best view of the stage.

The main hall, with 1,400 sq m of event space, has been integrated in the large space, almost 11 m high, over the submarine basin. The basin has been left in its original state, but kept out of sight beneath a newly laid concrete slab. Committed to the principle of universal space, the stage equipment is positioned on platforms along the sides of the pen and on two traversing gantries, 8 m high and 19 m wide, which can be moved along the full length of the hall. A 16 m-wide folding door permits the hall to be opened to the harbor. Visitors enter the hall from the landward side, via the former service corridor. This has been made accessible as a semi-public passage through the bunker. It also provides access to the roof of the bunker: ninety steps up a steel stair. Once on the roof, the tenants and visitors to Alvéole 14 can enjoy a terrace of 320 sq m. The main platform is roofed-over by a radome: a hemispherical dome that, until 2004, served as weather protection for NATO radar equipment at Berlin's Tempelhof Airport. It is made up of 298 triangular modules, with a transparent tensile membrane over an aluminum frame.

A novel method for cutting concrete with the assistance of diamond wire saws made it possible to cut through the exceedingly thick reinforced concrete. In order to establish access to the roof, the architects had a total of 470 cubic meters of concrete cut out and chiseled away, which is equivalent to a weight of approximately 1,000 tons.

Although the roof platform may, upon first glance, seem like a by-product of the conversion idea, the terrace and its 9 m high dome, which can be illuminated at night, are in reality an indispensable part of the overall concept: Without the striking dome, which shines over the harbor district like a bright, somewhat unfamiliar lantern in the darkness, lighting the way to Alvéole 14, the changes inside would not have been perceptible from the outside. Moreover, for the newly established use to assert itself visually in the face of the massive and hermetic nature of the built structure, it needs this strong sign, which is visible from afar. It was also essential to "break open" the concrete shell physically. The staircase to the roof and the roof terrace, with its view over the harbor, consequently appear to be a logical addition to the venues inside. FPJ

PROJECT DATA

Client City of Saint-Nazaire
Construction costs EUR 5.9 M (Alvéole 14), EUR 1.2 M (outdoor areas)
Usable floor area (net) 3,300 m² (interior), 2,270 m² (exterior)
GFA 5,250 m² (interior spaces)
Feature The opening in the concrete slab, which is several meters thick, was made using a diamond saw.
Completion 5/2007
Project management Hans-Michael Földeak
Location Submarine Base—Bay 14, Boulevard de la Légion d'Honneur, F-44600 Saint-Nazaire

Year of construction 1941

1000 ———————————— 2010

Conversion 2007

2000 ———————————— 2010

Construction costs   **Price per m²**
5.9 M €   1,788 €

15,000
10,000
5,000
0

CROSS SECTION THROUGH THE BUNKER'S 13ᵀᴴ AND 14ᵀᴴ CELLS

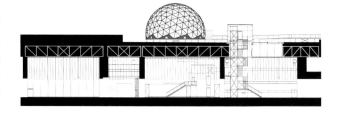

LONGITUDINAL SECTION THROUGH THE BUNKER'S 14ᵀᴴ CELL
with workshop and service corridor at the left and U-boat pen to the right

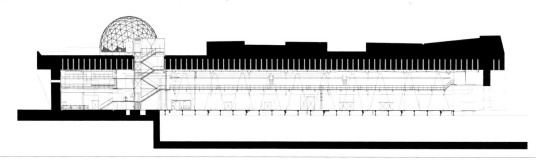

10 m

OLD & NEW

5   The vast hall in the U-boat pen during an event
6   The "street" with staircase leading up to VIP's administration offices and equipment rooms
7   Hundreds of fine spotlights illuminate the way in, along the former service corridor.

8 Installing technical equipment in the U-boat hall; stage lighting and sound equipment can be moved along the full length of the hall on traversing gantries.
9 The stage of VIP, a venue for events in the former workshop: view from the gallery
10 The bar on VIP's gallery

ZEHDENICK BRICKWORKS MUSEUM

ZEHDENICK-MILDENBERG (DE)
OLD Industrial — pp. 18, 132, 136, 148, 152
NEW Museum — pp. 32, 38, 42, 68, 178

— Duncan McCauley

1   [p. 161] An entrance structure at the side enables access to the upper floor of the landmarked ring oven.
2   Event hall on the upper floor of the ring oven. The old coal trolleys now serve as mobile buffets and cloak rooms.
3,4   Exhibition area demonstrating automated brick production
5   The exhibition "Bricks for Berlin" in kiln II explains the history of brick production in Zehdenick.
6   In the base of kiln II, visitors are acquainted with the phases of the firing process.

SPACE AS MEDIUM AND TESTAMENT

CONVERSION OF AN OLD BRICK FACTORY INTO A MUSEUM
DUNCAN MCCAULEY, BERLIN

Around 1900, a typical Berlin apartment building was comprised of a front building, two side wings, and a rear building, had forty apartments—and was constructed of 1.4 million clay bricks. Whoever wants to know where the billions of bricks came from that enabled the building boom during the *Gründerzeit* era in the capital of the German Empire will find the answer in Zehdenick, 50 km north of Berlin. In 1910, 625 million building bricks were fired here in fifty-seven continuous kilns known as "ring ovens." Manufacturing only ceased here in 1991. In the two remaining ovens and in the communist-era production building, Berlin architects Tom Duncan and Noel McCauley have created a brickworks museum. The site's history is explained in a space of roughly 5,000 m²—from the manual production of bricks to Hoffmann's 1858 invention of the ring oven and mechanized production in the post-war era.

The ring oven allowed bricks of uniform quality to be made in large quantities for the first time, within a relatively brief firing time: The unfired ("green") bricks were layered—arduously by hand—into dense stacks in the firing chamber. Then the workers fueled the chamber with coal and coke over the course of several days through openings in the ceiling. This meant that the bricks were first preheated, then fired at maximum temperature, and lastly allowed to cool down slowly.

The architects installed the exhibition *Bricks for Berlin* in one of the kilns. It presents the history of brick production in Zehdenick, once the source of bricks for nearby Berlin. The strikingly cubist, aubergine-colored exhibition furniture provides a distinct contrast to the yellow masonry of the firing chamber.

In the adjacent ring oven, Duncan and McCauley rely entirely on the impact of the space, supplemented only by a subtle collage of light and sounds: Where the unfired bricks were once piled up to the ceiling, visitors can wander along the 80 m long and 3 m high oval of the ring oven. A transparent brick that you receive at the entrance uses concealed electronics to illustrate the firing process: on your way along the firing channel, it changes its color to show how a brick would look at the respective stage of firing—the bricks were heated up to 980 °C until they glowed red. In this way, the former brickworks becomes a medium in which space, information, and the visitor's experience merge.

In addition to the exhibition, the 5.6 million euro project comprises redesign of the visitor center, restoration of two ring ovens (which are protected monuments), and the establishment of an event venue that can be rented for occasions all kinds. The architects integrated the 350 m² hall into the upper floor of the second ring oven as a "building inside a building." From here, the brickworkers used to heat the oven below. Duncan and McCauley have left the tracks for the trolleys that were used

PROJECT DATA
Client District of Oberhavel, Oranienburg
Construction costs EUR 5.6 M
Usable floor area approx. 5,000 m²
Gross volume approx. 44,550 m³
Feature The former firing chambers of the ring ovens have been converted to exhibition space.
Completion 4/2009
Project management Tom Duncan, Noel McCauley
Team Anuschka Müller, Katharina Bonhag, Lojang Soenario, Sandra Tebbe, Eva Maria Heinrich, Arno Kraehahn
Location Ziegelei 10, D–16792 Zehdenick-Mildenberg

Year of construction 1890

1000	2010

Conversion 2009

2000	2010

Construction costs **Price per m²**
5.6 M €   1,120 €

15,000
10,000
5,000
0

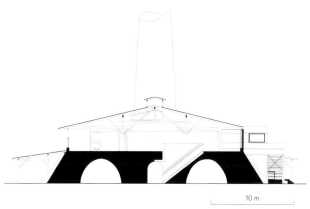

CROSS SECTION THROUGH KILN II

10 m

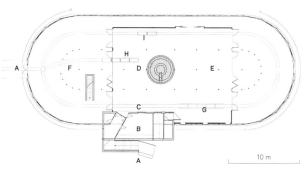

UPPER FLOOR PLAN after conversion to an event hall

10 m

A Entrance	D Smokestack	H Cloak-room lorries
B Annex	E Stage	I Bar lorries
C 'Building inside a building'	F Existing structure	
	G Buffet lorries	

for transporting coal in the floor of the hall. Nine restored trolleys have been converted into a buffet and cloak rooms. In summer, the entire hall tends to be in use, whereas in winter wide folding doors allow functions to be concentrated in the inner hall surrounding the smokestack. The external appearance of the oven with its conical brick base, wood additions, and imposing smokestack remains unchanged. Only the addition at the side, in which the stairway and a vestibule are found, allude to the newly created event hall on the upper level.

Tom Duncan and Noel McCauley have designed every detail of the museum themselves, beginning with the building and including the exhibition design, the illustrations, and the films shown. The architects' concept assembles relics of production, visual documents, descriptions, production sounds, and film material within the landmark building into a total museum experience. FPJ

Upper floor of the ring oven before renovation

SELEXYZ

MAASTRICHT (NL)
OLD Church
NEW Retail —p. 50

DOMINICANEN
BOOKSTORE
— Merkx + Girod Architecten

The Gothic Dominican church stands on a small, secluded square in the historical district of Maastricht. Finished in 1294, the building originally served as a monastery church for the Dominican Order until the Napoleonic troops descended in 1796 and the monastery was disbanded in the course of secularization. In the following centuries, the church—with neither tower nor transept—served a multitude of different functions, none of which were very dignified. To begin with, it was used as a stable for the cavalry, then as a warehouse for the municipal fire department; later it became an event hall for boxing matches, cactus shows, and carnival parties and most recently, it even had to serve as a garage for bicycles. The many tourists who visited Maastricht ignored it in any case, because there are many better-preserved churches to see at more prominent locations in the city.

LITERATURE INSTEAD OF LITURGY

BOOKSTORE IN A GOTHIC CHURCH IN MAASTRICHT
MERKX + GIROD ARCHITECTEN, AMSTERDAM

ANNEKE BOKERN

When Merkx + Girod received the commission for the renovation, they had already furnished one store in Almere and another in The Hague, both for the same client—the bookstore chain BGN. The point of departure for the design of the Dominican church was the architects' desire not to obstruct or disguise the church space, and to retain its sacral atmosphere. The only question was: How do you install a bookstore with a selection of 30,000 titles without impairing the existing spatial quality? Only 550 sq m of floor area were available in the church overall, but nearly 1,000 sq m were needed. The client initially suggested the insertion of additional levels into the church, in the form of bridges. But the landmarks preservation council had prohibited any measures that might damage the church walls. All built-in fixtures had to be removable in such a way as to permit the building to be returned to its previous state.

A solution was then found, in the form of an oversized, walk-in bookcase. Merkx + Girod have positioned it in a decentralized position in the right half of the church, so that a mural from the year 1337 on the left wall of the church remains visible. Overall, the steel structure is 30 m long by 18 m high and has three levels, each with three rows of shelves in varying formats. It stands freely in order to separate the piers, central nave and the side aisle from one another, and accommodates two workplaces in addition to the bookshelves. Customers use a stair that leads alternately between colorful book spines or the white trimmed edges of the books' pages. Once at the top, visitors get a view of the church to that makes its magnitude clear for the first time, and find themselves closer to the column capitals and painted vaults than ever before.

The remaining interventions in the church are modest in comparison to the stunning bookcase. For displaying bestsellers and special offers, Merkx + Girod have furnished the church with the same system used in their stores in The Hague and Almere. Here in Maastricht, though, they have consciously kept the wall-mounted shelving and the display tables low, so that they barely infringe on the impact of the space. With its half-round form and more comfortable dimensions, the chancel presented the best location for the café. Its focus is a cruciform reading table. The toilets and equipment are located in the basement underneath the chancel. All the remaining technical fixtures, the underfloor heating, and the few gravestones that have survived the centuries have been integrated into the new concrete floor. It replaces an unattractive surface of paving stones, which dates from the days when the church was used as a bicycle garage.

Since the full height of the space and the view from the entrance toward the chancel have both been retained, the Dominican church has forfeited little of its spatial effect, despite the installation. At the same time, the giant bookcase provides new perspectives and establishes an effective contrast within the church—between industrial and handcrafted elements, new and old, smooth and rough, heavy and fine. Moreover, owing to the spectacular installation, not only customers of the bookstore, but also tourists are suddenly attracted to this once-forgotten Dominican church.

PLAN OF THE GROUND FLOOR OF THE DOMINICAN CHURCH

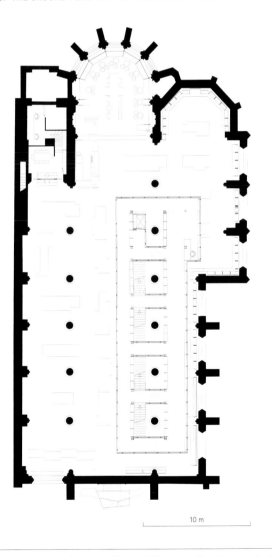

10 m

Ever since the Amsterdam office of Merkx + Girod converted the church into a bookstore, however, the hidden location has proven to be advantageous: when you step through the portal from the small square, you find yourself in an unexpectedly large church. No less astonishing than the jump in scale between the surrounding cityscape and the church's interior is the gigantic, multi-level steel bookcase that presides against the church's right wall.

PROJECT DATA
Client BGN Boekhandelsgroep Nederland, Houten
Construction costs EUR 1.6 M
Usable floor area 1,200 m²
Feature Revitalization and urban reappropriation of a previously abandoned ecclesiastical building through dignified commercial reuse
Completion 1/2007
Project management/team Evelyne Merkx, Patrice Girod, Bert de Munnik, Abbie Steinhauser, Pim Houben, Josje Kuiper, Ramon Wijsman, Ruben Bus
Location Dominikanerkerkstraat 1, NL–6211 CZ Maastricht

Year of construction 1294

1000   2010

Conversion 2008

2000   2010

Construction costs   **Price per m²**
1.6 M €   1,333 €

15,000

10,000

5,000

0

1 [p.164] Church as bookstore: view toward the apse
2 In the apse of the former house of worship, there is now space for a coffee break.
3 Whereas only low display tables have been placed in the northern half of the nave, the large, walk-through steel bookcase fills its southern side.
4 Who can concentrate on books with a view like this? Perspective from the 2nd level of the bookcase
5 From outside, no one would guess the church building's new function.

ST. JOSEPH'S HOUSE AT WALDSASSEN ABBEY

WALDSASSEN (DE)
OLD Farm building —p. 172
NEW Restaurant/Bar —pp. 28, 92, 136, 156
Hotel —p. 136

— Brückner & Brückner Architekten

1   The refurbished gable end of the old malthouse, freed of all additions
2   The square in front of the abbey church from 1704; in the background on the right is St. Joseph's House.
3   Cistercian nuns on the renovated square
4   The modern entry structure in front of the angled malthouse building

THE ART OF EXTENSION

CONVERSION OF A MALTHOUSE INTO A CULTURAL AND COMMUNITY CENTER FOR WALDSASSEN ABBEY
BRÜCKNER & BRÜCKNER ARCHITEKTEN, TIRSCHENREUTH/WÜRZBURG

Actually, Waldsassen Abbey's old malthouse was to have been demolished more than two hundred years ago. During the eighteenth century, the entire abbey complex was rebuilt in the Baroque style and substantially enlarged according to plans of the prominent master church builders Abraham Leuthner and the Dientzenhofer brothers. In 1704, the magnificent Baroque church was consecrated, and in 1727 the library was completed.

Built in the middle of the fifteenth century, since when it had been used as a stable, smokehouse, malt kiln, storage cellar, and dormitory, the austere outbuilding with a tall gable was the last relic of the medieval abbey complex and was due to be demolished. Then, suddenly, work came to a halt; Napoleon's troops swept across Europe, the abbey was secularized, and the books from its renowned library were dispersed. So the last medieval outbuilding remained in place, flanking the forecourt of the abbey's basilica. Today, it marks the urban conjunction of the basilica, the adjacent abbey complex, and the town. Over the centuries, though, the building not only became dilapidated, but numerous renovations and additions gave it the appearance of an unattractive conglomeration.

In the mid-1990s, the Cistercian convent experienced a new beginning: The convent, at times ageing and in danger of extinction, was rejuvenated by the addition of young, new nuns. This euphoric mood spurred on the plans for a general refurbishment—the first since Baroque times. In the course of the planning, the potential of the vacant malthouse was also discovered. It was resolved to make the desolate, but nevertheless oldest remaining building of the complex into a beachhead between the convent and the secular world, as a cultural center, shelter, guesthouse, and shop.

At first, the architects investigated the built substance, identified the different historical layers and searched underneath the stucco for structural junctions, old openings, and traces of paint. Ultimately, numerous partition walls, as well as additions and sheds, were demolished, especially along the inner sides of the angled building. In place of the demolished extensions, there is a newly appended circulation zone with pergola-like protruding buttresses: This belatedly implements the abandoned Baroque plans for renewing this part of the abbey. The old malthouse and the abbey wing, built in the Baroque style, which adjoins it at one end, share the same alignment, but the eighteenth-century building is about four meters wider than the earlier one. This difference in scale, evident in the more stately dimensions of the Baroque abbey, has been transferred in a contemporary style to the service building by Brückner & Brückner. The struts of the appended access gallery were constructed of steel, after first considering and then rejecting both wood and concrete. This new boundary of the western basilica square defines the abbey's forecourt as an entity in the urban context.

When they had cleared the superfluous material away, the architects found a solid core to the existing building that was worthy of preservation. Nevertheless, without fundamental intervention and also some cunning manipulation, the existing substance would not have fulfilled the requirements for future use. One of these pieces of manipulation concerned the groined vault on the ground floor, which was much too low for use as the abbey's restaurant. The architects decided to replace the floor and to extend the original piers of the groined vault downward by inserting custom-made granite plinths. No-one who enters the restaurant today would suspect that it would once hardly have been possible to stand upright beneath the original vault. Here, as elsewhere, the architects have consciously restricted themselves to materials available when the building was initially erected—fieldstones, bricks, and wood—but they have used them with a decidedly modern architectural language. The goal of exploiting the historical attic space as new usable floor area made extensive rehabilitation measures necessary in the roof structure.

The refurbished building is divided into three parts: the ground floor and basement are used semi-publicly for the shop, restaurant, and reception; the upper floor accommodates the guest rooms, and the attic has been fitted out as a versatile seminar area. FPJ

PROJECT DATA
Client Kloster Waldsassen GmbH, Reverend Mother Superior, Sr. M. Laetitia Fech
Construction costs EUR 5.7 M (cost groups 300–600)
Primary usable area 2,100 m²
GFA approx. 3,250 m²
Gross volume approx. 11,100 m³
Feature All timber removed (from demolition of the outbuildings and some joist floors) was reused in the course of refurbishing the historical roof frame.
Completion 9/2008
Project management Stephanie Reichl
Location Brauhausstraße 1, 3, D–95652 Waldsassen

Year of construction 1450

1000		2010

Conversion 2009

2000		2010

Construction costs	Price per m²
5.7 M €	2,714 €

SECTION THROUGH THE BUILDING WITH ADDITION

10 m

PLAN OF THE HOTEL first upper floor

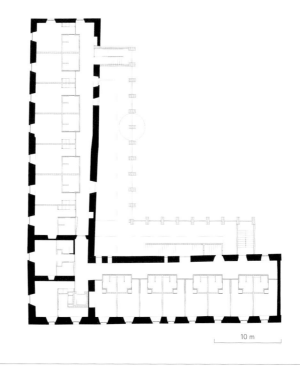

10 m

5,9 The ground floor level after its conversion into a restaurant. Fig. 5 shows the vaults during renovation; in order to gain the needed ceiling height, the floor level was lowered and the pillar was elongated with a granite plinth.

6 The L-shaped addition connects all areas of the building with each other.
7 A guest room
8 Multi-purpose room in the attic

SCHWEINFURT (DE)
OLD Farm building — p. 168
NEW Library — p. 178

LIBRARY
IN EBRACHER HOF
— Bruno Fioretti Marquez Architekten

GAME OF REFLECTIONS

CONVERSION OF A TITHE BARN INTO A PUBLIC LIBRARY
BRUNO FIORETTI MARQUEZ
ARCHITEKTEN, BERLIN

SIMONE JUNG

The Zehntscheune in Schweinfurt was built in 1431 to serve the nearby Cistercian monastery at Ebrach. Since then it has experienced some turbulent times. As a center for the production of ball bearings, the small city in northern Franconia became the target of Allied bomber squadrons in the Second World War. The Zehntscheune, a medieval tithe barn with a stepped gable, was one of only a few buildings from the Middle Ages to survive the raids. A little later, in the early 1960s, the adjacent monastery garden and a large portion of the historical city fortifications along the River Main were lost to a new riverside road. The tithe barn, once closely surrounded by buildings, was suddenly exposed on two sides, leaving it somewhat lost next to the bridge over the Main that carries the southern access road to the city.

When the city of Schweinfurt held a competition for the conversion of the vacant building into a municipal library, the winning entry was the only one to propose that the additional spaces required should be put underground, so as to leave the forecourt free and preserve the external appearance of the ensemble.

In this design, by Bruno Fioretti Marquez Architekten of Berlin, a spacious, subterranean level surrounding the six-hundred-year-old remnants accommodates most of the library. The involuntary and arbitrary clearance of the 1960s is thus succeeded by a well calculated one. Freeing old building parts and singling them out admittedly has a tradition as an aesthetic option when building within the existing fabric: Carlo Scarpa left his mark on this strategy and many applied it subsequently, for instance Egon Eiermann, with his conversion of Berlin's Kaiser Wilhelm Memorial Church into an aestheticized ruin. By burrowing underground, Bruno Fioretti Marquez explore the potential of this approach afresh: The Zehntscheune is not only freed at parts that were visible thus far, but also beneath the earth, to its foundations.

Bruno Fioretti Marquez were particularly interested in giving the building's historical substance a new appeal: Thus, the massive timber joists within the building have been refurbished and painted white; the base of the building, with its carefully repaired fieldstone masonry, has been highlighted within a bright and expansive hall. On the square, only a glass enclosure provides evidence of the enlargement below. A rigorous, 33 meter-long glass prism illuminates the lower level during the day and allows the library to shine out toward the city at night. As a glass front, it shields the forecourt from the heavy traffic on the riverside road. A new main entrance in one of the gable walls provides access to the public library from the forecourt. From the entrance, a stair first leads down to the lower level. From a slightly raised position, you can let your gaze wander over the high, long space and the seemingly endless wall of books on the left, before you explore it along a gently sloping ramp that runs beside the masonry foundations. Aestheticization, combined with the greatest possible degree of refinement, account for the idiosyncratic fascination of the subterranean space. Nothing except the books distracts from the triad of quarried stone, exposed concrete, and oak wood. Ventilation ducts and other technical equipment have been concealed behind shelves and within inconspicuous shadow joints, just as the lighting has been—not a single light fixture is to be seen in the entire

hall. Bruno Fioretti Marquez have recognized that the aesthetic potential of such a contrast of old and new, smooth and rough, only comes into its own when the new elements respond to the characteristics of the existing fabric.

To begin with, the bend in the long bookcase wall attracts attention: In its last third, the concrete frame—and with it the wood shelving—slopes gently downward. Although the solid oak columns extend perpendicularly to the ceiling, this does not run horizontally and so they, too, are inclined. Here, the concept incorporates imperfection: the architects did not want to brusquely contrast the deformations of the existing building—which have appeared over the course of centuries—with the millimeter-precise perfection of modern building techniques. In what they have added, they respond to the irregularities of the existing building "with gentle breaks in the strict orthogonal regularity," as architecture critic Falk Jaeger expresses it.

It was a stroke of luck that the architects also received the commission to build the new customs office nearby. They have not wasted the chance to allow both buildings to enter a dialogue. The official building, constructed at a right angle to the facade of the Zehntscheune, has large-format, protruding windows. As with a puzzle, each one of the sixteen windows reflects a different part of the Zehntscheune—these fragmentary images overlap and repeat themselves, because some of the windows placed asymmetrically in the concrete facade are tilted out from the front toward the library by a few degrees. This game of reflection in the openings is like interpreting the past—from every perspective it looks somewhat different.

PROJECT DATA

Client City of Schweinfurt, represented by the building department
Construction costs EUR 6.6 M
Usable floor area 1,281 m²
GFA 2,498 m²
Gross volume 11,282 m³
Completion 4/2007
Project management/team Wieland Vajen, Simone Skiba
Location Brückenstraße 29, D–97421 Schweinfurt

Year of construction 1431

Conversion 2007

Construction costs | Price per m²
6.6 M € | 5,152 €

LONGITUDINAL SECTION THROUGH THE CONVERTED TITHE BARN

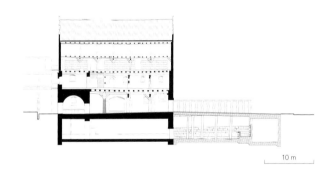

10 m

PLAN OF LOWER LEVEL

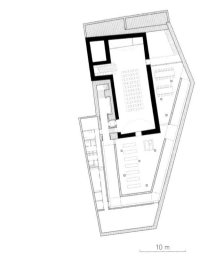

10 m

1   [p.172]
1   [p.172] Disjointed views of the old stepped gable are reflected in the large format windows of the neighboring main customs office.
2   Overall view: The converted tithe barn and the new main customs office (left) seen from the bank of the River Main
3   The long glass form of the lantern shields the forecourt from the adjacent street and illuminates the subterranean spaces of the library.
4   Discreet: the added egress stair tower extends the profile of the adjoining gable, abstracted in exposed concrete.
5   Side view of the refurbished barn; in the background is the Georg Schäfer Museum by Volker Staab.

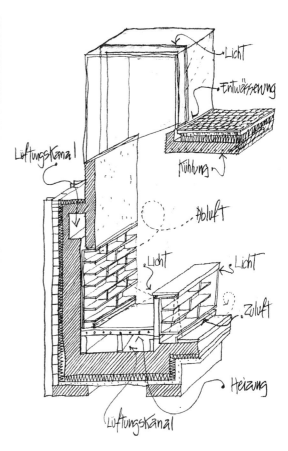

7|8

6   The reading room on the lower level
7   The main space of the library on the lower level, with the exposed cellar wall of the tithe barn at the left
8   Sectional sketch of the subterranean room, showing concealed technical services
9   View of the lower level
10   On the upper level, light painted oak columns contrast with dark furnishings.

TERRA MINERALIA IN FREUDENSTEIN CASTLE

FREIBERG (DE)
OLD Warehouse —pp. 22, 68
NEW Library —p. 172
Museum —pp. 32, 38, 42, 68, 161

— AFF Architekten

1

In 2005, the young office of AFF Architekten was awarded first place in a Europe-wide competition for the rehabilitation and renovation of Freudenstein Castle in the Saxonian town of Freiberg.

The former royal residence is an enclosed, four-wing ensemble on the edge of the town center. Constructed in 1168 by the Wettin dynasty as a fortress to guard the silver mines, it was enlarged in grand manner during the Renaissance. During the

RENAISSANCE OF THE CABINET OF CURIOSITIES

MINERALOGICAL COLLECTION IN A CONVERTED CASTLE
AFF ARCHITEKTEN, BERLIN / CHEMNITZ

the architects to convince the client and future users to retain the timber construction, at least in an exemplary area three bays wide. Elsewhere, too, AFF Architekten have recognized the vestiges of Freudenstein's checkered history as aesthetic potential for the new use.

On the ground floor, where the "treasure chamber" of the minerals collection was to be located, the ceiling vaults were blackened with soot, presumably from use as the kitchen of

Seven Years' War, however, its interior was mostly destroyed. A period of decline followed. At the end of the eighteenth century, the ownerless castle was converted into a depot, and it served as a granary until 1979.

By the time that the city of Freiberg decided to use the ensemble for cultural purposes, its former splendor was evident in little more

PROJECT DATA
Client Freudenstein Castle: City of Freiberg
Client Mineralogical Collection: Sächsisches Immobilien- und Baumanagement, Chemnitz
Construction costs Freudenstein-Castle: EUR 21.4 M, Mineralogical Collection: EUR 3.35 M
Usable floor area (Mineralogical Collection): 1,500 m²
GFA 16,450 m² (entire building)
Gross volume 59,120 m³
Completion 1/2008
Project management Martin Fröhlich, Sven Fröhlich, Alexander Georgi
Location Freudenstein Castle, D-09596 Freiberg

Year of construction 1577

1000 ———————————————— 2010

Conversion 2006-2008

2000 ———————————————— 2010

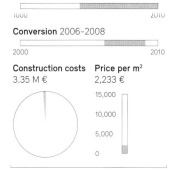

Construction costs **Price per m²**
3.35 M €   2,233 €

15,000
10,000
5,000
0

than some stairways and the Renaissance gable over the gateway from the bridge. It was decided that the castle should house the mineralogical collection of Freiberg technical college and mining academy—considerably increased through a donation—and the Saxonian Mining Archives, one of the most extensive collections of mining documents in Europe.

The archive and the collection are entered through a shared entrance in the palace courtyard. Not least for reasons of conservation, the archive has been inserted as a self-contained "archive corpus"—with a secondary enclosure—in the so-called "church wing." The mineral collection, known as Terra Mineralia, has moved in to three floors of the diagonally adjacent northeast wing.

Here, in contrast to the archive's wing, nothing stood in the way of direct interconnection with the historical architecture. The existing elements consisted of , on the one hand, remains of Renaissance features and, on the other hand, elements from the eighteenth century, in particular the formidable depot floors, constructed from solid oak beams, each level of which corresponds to half the height of a Renaissance story. As in the archive's wing, these storage levels were initially to be removed entirely. It took considerable negotiating by

a military hospital in Napoleonic times. The architect's desire to keep these traces of history from vanishing beneath fresh plaster surprised the client. The client finally consented to this, though, so the treasures now lie beneath a strangely marbled, partly black and partly bluish shimmering vault, which almost seems like a chapel with its dramatic lighting, but also conjures up thoughts of an artificial cave. A more fitting and subtly complementary backdrop for the enigmatic beauty of the minerals and crystals is hardly imaginable. The spirit of its new contents pervades the castle, in the form of its architectural renovation. Particularly the motif of mineral inclusions—which are accumulations of crystals in the cavity spaces of rocks—has been adopted repeatedly by the architects: surrounded by a rough, monochromatic concrete shell, you encounter built-in elements in glowing colors: violet, bright yellow and deep green. In the rooms for the collection itself, care has been taken to use an exceedingly restrained color scheme—all display cases and additional exhibition furniture are black. The visitor's attention is directed entirely towards the miniature world of crystals.

The architects have not sought contrasts or demonstrative breaks between old and new: Although the new elements come very effectively to the fore, elsewhere the existing building and the contemporary addition are imperceptibly blurred. Some rooms have been restored so as to re-establish the spatial layout of the Renaissance. Because the architects were able to design all the furniture and built-in fixtures in a manner consistent with their cautious restoration of the existing fabric, exhibition spaces of great expressiveness and formal unity have been created. Their scenic effect is restrained and simultaneously suggestive. Unavoidably, a "cabinet of curiosities" comes to mind. This term for royal collections of the Renaissance, the forerunners of modern museums, refers to the fascination of the viewer with the marvels that could be found in them. Seen in this context, Terra Mineralia is indeed a modern cabinet of curiosities. Oddly branching crystal needles, which have been brought to light from cavities deep within the earth, compete with solemnly barren spaces for the amazement of the visitors, and with a supporting structure that looks as if it could bear an entire mountain. Content and form are completely in harmony. FPJ

CROSS SECTION AND LONGITUDINAL SECTION THROUGH THE COLLECTION WING

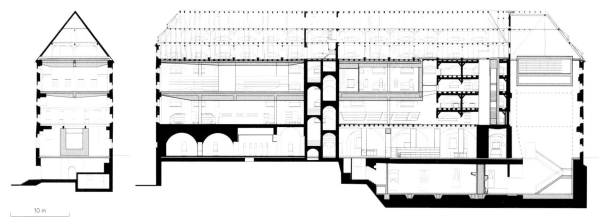

10 m

5|6

7

8

1   [p.178] The castle's courtyard after renovation: the archive wing is at the center, the collection wing is on the left. The new entrance in the courtyard provides joint access to both.
2,3   The new entrance
4   The renovated Renaissance stairway with new lighting
5,6   In the 2nd and 3rd floors, the architects were able to preserve the intermediate floors that had been added when the castle was converted to a warehouse. The exhibition display cases and the other interior fittings used for the collection were also designed by AFF Architects.
7,8   Dark exhibition furniture contrasts with historical traces: the treasury in the lower level vaults

LANDMARKED BUILDINGS —AUTHENTIC DESPITE NEW FUNCTIONS

— Three Refurbishment Stories from Baroque to Bauhaus

Building work to existing buildings has long overtaken new construction activity and is set to increase even further. This broadens the opportunities for giving buildings of special historic interest— landmarked buildings—a new function and hence a new lease of life, irrespective of whether they are in use or vacant, and thereby preserving them for our cities.

WINFRIED BRENNE / FRANZ JASCHKE

With the three examples of our work illustrated below, we hope to demonstrate that work to buildings with landmark status tends to be particularly successful where the specific qualities of a building are recognized and utilized, and that while exercising all due respect and care in dealing with the existing substance, the design should not be compromised.

We also describe the methodology and approach, which always varies with the type of building, its current condition and the design brief, and which needs to be considered in the context, on the one hand, of the requirements for protection as a historic building and, on the other hand, the need for a modern energy-efficient building envelope and appropriate building services.

I
REFURBISHMENT OF THE BERLIN ARMORY TO HOUSE THE PERMANENT EXHIBITION OF THE GERMAN HISTORICAL MUSEUM: NICHES AS A DESIGN IDEA
—

In 1998 the Federal Government organized an architectural project competition for the conversion of the Armory (Zeughaus) to create space for the future permanent exhibition of the German Historical Museum. There was an initial design with suspended ceilings and raised floors for accommodating services installations, which had been rejected by the client as unsuitable, because it would have changed the interior effect of the

Armory very substantially and would have robbed the building of its character. The landmark preservation office also stipulated that such extensive changes should be avoided and requested that the different architectural legacies and traces of its three hundred year history be made visible, bearing in mind that this is one of the last monumental Baroque buildings in the heart of Berlin.

The competition brief called for the building to be brought up to the standard required in modern museums: It included technical requirements such as full air conditioning of the exhibition rooms and the creation of varied lighting scenarios, as well as all safety requirements regarding fire safety, smoke extraction, burglary protection etc.

In addition, the internal organization of the building had to be optimized. The objective was to increase the exhibition areas, to create additional space for archives and storage, to improve visitor circulation, and to optimize access for the disabled.

For the conceptual approach it was essential to explore, get to know, and understand the building with respect to its structure, geometry, and sense of scale.

But above all, this understanding of the building needed to be underpinned by lateral thinking and by questioning seemingly obvious solutions. This was the only way

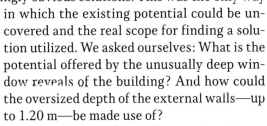

in which the existing potential could be uncovered and the real scope for finding a solution utilized. We asked ourselves: What is the potential offered by the unusually deep window reveals of the building? And how could the oversized depth of the external walls—up to 1.20 m—be made use of?

This questioning led us to discover the niche as a future location for a decentralized air conditioning system. Our services engineer Matthias Schuler (Transsolar) welcomed this idea, which was soon worked out in greater detail after a joint design session.[1] It corresponds to the trend towards decentralization of building services. In addition, this design approach gave us the opportunity to reduce interference with the existing building fabric and find a sensitive way of adding elements in a modest, but nevertheless confident, architectural expression, without detracting from the overall historical context of the building. The fact that the idea was not accepted without some supporting arguments is evident from the jury's assessment: "...The proposed ventilation system is considered to be innovative although the full details of its functional capability cannot yet be ascertained ..."

We were therefore asked by the client to check our idea in a step-by-step process and to progress it beyond the

experimental stage, something we also wanted to clarify for our own peace of mind. The client made it a fundamental condition of implementation of the design that the planned niche-based air-conditioning unit be developed to full serial production readiness (rather than just a prototype).

The very vivid visualization of the thermal and airflow simulation produced by the design team was not considered to be sufficient.

In response, the manufacturer of the air-conditioning units produced a 1:1 glass mock-up, including all technical components, to replicate the situation in the

CONVERSION OF THE BERLIN ARMORY FOR THE GERMAN HISTORICAL MUSEUM Diagram of decentralized air-conditioning unit at the window niches

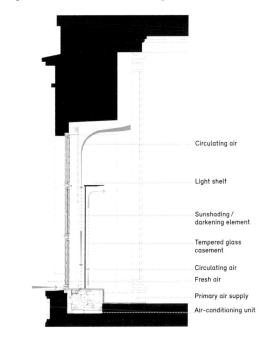

Circulating air

Light shelf

Sunshading / darkening element

Tempered glass casement

Circulating air
Fresh air

Primary air supply

Air-conditioning unit

building. Thanks to the glazing it was possible to produce a video that clearly demonstrated the air flow and effectiveness of the units, using fog trials. However, even that video failed to fully convince the client.

We responded by creating a test room for a 1:1 field test in the Armory by running a partition wall along the full 25 m depth of the building. In this room, summer and winter conditions were tested for a whole year, taking into consideration all relevant factors, such as sudden increases in humidity level.

During that one year test phase, we were able to prove that it was possible to create a stable room climate without short-term fluctuations, using these niche units installed in the facades opposite each other; furthermore, by then the manufacturer had developed the units to full serial production readiness. They had already been on the market as facade units for the heating

1   The German Historical Museum in the Berlin Armory: staircase leading to first upper floor

2   View of window niches after their modification to accommodate decentralized air-conditioning units: the technical components are concealed by an inconspicuous window sill.

and cooling of room air, but the humidification and dehumidification function was new.

While the development of the technical performance of the units was progressing, we worked on designing the overall unit with its details. In order to be able to add the necessary new elements, such as the casing of the units, the air-and-light-channeling glass elements and the light shelf, we designed a component with a shape that fits naturally into place; so much so that, at the opening ceremony of the building, the State Historic Preservation Officer characterized the item as "the art of least interference."

The system has now been in operation for six years and has passed the test of time not only with regard to architectural design and historic buildings, but also in terms of economy in operation: The decentralized air-conditioning system meant that the services centre could be significantly reduced in size and that large air ducts for distributing conditioned air could be dispensed with, which in turn resulted in significant cost and energy savings. It also enabled us to expand the exhibition areas and create additional space for archives on the attic floor, while preserving the authentic interior appearance of the building.

II
REFURBISHMENT OF THE BUILDING OF THE FORMER FEDERAL SCHOOL OF THE ADGB IN BERNAU FOR USE AS A BOARDING SCHOOL OF THE IIWK BERLIN

—

The former Federal School of the General German Confederation of Trade Unions (ADGB) in Bernau was built in 1929-1930 to a design by Hannes Meyer, which had won first prize in a competition organized by the

ADGB. The objective of the design was to create a large-scale architectural expression of the idea of teaching under a progressive, social and functional teaching paradigm. A distinct pattern language of materials defines the interior and exterior design of the building blocks. The complex is one of the most prominent historic Bauhaus projects outside the Bauhaus city of Dessau. The school, with its small lake, was originally conceived in the context of an open woodland area, but is today surrounded by other schools and educational buildings, which together form a campus-style development.

Following the end of communist rule in East Germany, the building complex, which during the preceding period had also been used as a trade union school, remained unoccupied for several years, leading inevitably to damage to the building fabric. The need to find a suitable user for the building grew from year to year.

It must be considered a major stroke of luck for this historic building that the Brandenburg Ministry of Higher Education, Research and Culture, the Brandenburg State Historic Preservation Office and the Berlin Chamber of Skilled Crafts and Small Businesses (HWK Berlin) agreed to house the HWK boarding school, which had initially been planned as a new building on the adjacent educational campus, in the former Federal School. This meant that the complex was once again utilized in almost the same way as before.

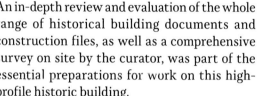

For us, the challenge was to find a workable compromise between current standards for the accommodation of trainees and a revitalization of the original architecture. The best possible modern standards had to be incorporated in this historic building, while also refurbishing its physical building fabric. The once-distinct features on the entrance side, with its three high chimneys, had changed beyond recognition and the eventful history of the building had left heavy scars on its interior and exterior.

An in-depth review and evaluation of the whole range of historical building documents and construction files, as well as a comprehensive survey on site by the curator, was part of the essential preparations for work on this high-profile historic building.

In this case, the special task was to record the quality and condition of the materials that determine the stylistic features of the building and to determine which of those features could be preserved with due respect to the period of their creation. The many layers of construction work, in the form of conversions and extensions, made it impossible to analyze the original structure fully until building work had started, cladding had been removed, and solid construction parts had been demolished, meaning that the survey could not be completed fully until building work had progressed to an advanced stage.

This also meant that the design and the landmarked-building concept had to be adapted and refined as work progressed, requiring a high degree of flexibility on the part of all those involved, as well as a permanent process of coordination. As a result, many decisions had to be made on site as the work progressed.

This process, which was not a simple one, was also supported by the client and the historic buildings officer as important decision-makers; special praise needs to

3 [p. 186] An aerial view from the time of construction shows how Hannes Meyer set the buildings into the landscape.

4 [p. 186] Condition of the student residence prior to refurbishment: the wooden windows installed during the communist period considerably changed the effect of the facade.

5 [p. 186] The wing with student accommodation after refurbishment: the clear contrast between the slender window sections and the brick facade has been re-established.

6 [p. 186] The solarium between the dining hall and student accommodations is formed as a quadrant. The architects reconstructed the addition based on original plans and foundation remnants.

go to the building control and fire safety service officers, without whom much of the rediscovered authenticity of the historic building would have been irretrievably lost. While aiming to maintain the required safety standards, they used their discretion to make best use of locally available resources and to interpret the relevant regulations favorably, granting exceptions and exemptions in order to be able to preserve special features of the original construction.

One such case was the second escape stair, which is normally required in residential buildings. In order to avoid this element with its disruptive design aspects, which was not part of the historic building, we agreed to create optimum conditions for the fire service to reach each individual room by ladder: Laminated safety glass was used in the windows. In order to facilitate breaking the glass in an emergency, each window is equipped with an emergency hammer, such as are in use in public transport vehicles. Likewise, an arrangement was worked out to allow an exception regarding the rather low height of the window sills and stair railings, which was that all new arrivals had to be instructed as to the special features of the building.

With respect to the building's authenticity, its originality, and the question as to which of the existing building components were to be demolished in order to re-establish the original architecture by Meyer and Wittwer, and which of those components could be integrated into the overall concept, we opted to assess each item on its merits: For instance, the high-quality additions dating from the 1950s particularly in the main building- the red brickwork of which can easily be distinguished from the yellow facades of the original building, were

WINFRIED BRENNE
—

Dipl.-Ing. Winfried Brenne, Architekt BDA/DWB (Deutscher Werkbund), born in 1942, studied architecture in Wuppertal and Berlin. Since 1978, he has been a freelance architect in Berlin, working mainly on housing developments, ecological construction, and building within the existing fabric. A primary focus of his work consists of condition surveys and the restoration of buildings and ensembles of the New Objectivity movement, including the Hufeneisensiedlung in Britz and other Berlin housing estates of the 1920s. In addition, the restoration of prominent individual landmarks, such as the Muche-Schlemmer House in Dessau and the former Federal School of the German Trade Union Alliance (ADGB) in Bernau, as well as Schloss Charlottenburg and the German Historical Museum in Berlin. Since 1993, he has been a member of Docomomo Germany; since 2000 of the German national committee of ICOMOS, and since 2006 of the Akademie der Künste, Berlin. He has published numerous books and articles, notably on the work of Bruno Taut, and has received awards at home and abroad.

FRANZ JASCHKE
—

Dipl.-Ing. Franz Jaschke, Architekt DWB, born 1955 in Meschede, North Rhine-Westphalia, put down roots in 1975 in Berlin (Kreuzberg). He gained his Diploma in Architecture from the TU Berlin in 1981 (Climate-conscious construction). Since 1983, he has collaborated with Winfried Brenne in different constellations. He founded BRENNE GmbH in 2001 in conjunction with the commission to restore the former ADGB Federal School in Bernau originally designed by Hannes Meyer and Hans Wittwer (project awarded the WMF/Knoll Modernism Prize in 2008, among others). Among the projects which he has been responsible for are the Bruno Taut residence in Dahlewitz, the restoration of the former Reichstagspräsidentenpalais in Berlin, the Neuer Flügel of Schloss Charlottenburg, a pilot project for ecological housing in Berlin-Pankow and a research project on life-cycle assessment and environmental accounting. He is a founding member of Docomomo Germany and belongs to the Deutscher Werkbund Berlin.

www.brenne-architekten.de

retained. On the other hand, the facades of the residential buildings and the school building itself—with their yellow facing brick and flush fair-faced structural concrete elements—were exposed so as to re-establish the historic elevations, complete with such traces of use that time had left.

The internal room layout in the residential buildings was not changed, although of course consideration had to be given to the fact that today's standards for the accommodation of trainees are different from those of eighty years ago. Instead of the original washbasin in each two-bed room and a shared bathroom on each corridor, each room was fitted with its own small shower room.

Rooms and corridors were decorated in the color-coded scheme developed by Hannes Meyer so that today, the occupants can again orientate and identify themselves with the red, yellow, green, or blue house. Although the remains of interior finishes and colors found during the initial survey were rather patchy, a full reconstruction was possible, owing to later discoveries of remains of wallpaper and color finishes found behind skirting and in similar places.

In the residential buildings, we found wooden, double-casement windows, which had been installed during the communist period, probably because of the then poor state of the original windows, combined with energy conservation considerations. In our design these were replaced by steel windows, which it was possible to reconstruct true to the historic pattern, on the basis of some original windows still in existence. At the time of the conversion, from 2002 to 2006, thermally separated profiles were not yet commercially available for this situation, such as those that are currently being developed and which will soon be available in serial production.

As cold bridges exist at the windows, as well as at the solid concrete lintels over the windows—which

7   [p. 186] The glazed ceiling and its elegant supports give a light, airy atmosphere to the dining hall; shown here after refurbishment.

8   Reconstructed steel window in a student apartment: An emergency hammer has now been provided for each window, making it possible to dispense with the mandatory second escape route.

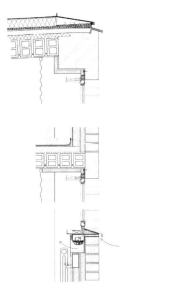

SECTION THROUGH WALL AND WINDOW IN THE STUDENT RESIDENCE
The internal insulation lining the window lintels and the double-glazing used in the new steel windows significantly improve the energy-conserving characteristics of the building.

originally were not insulated—the windows were fitted with double glazing and internal insulation was installed where required. In addition, a simple ventilation concept was developed as compensation, using the extractor fan in the bathroom to reduce excessive humidity, which could otherwise form condensate in those areas.

Sensitive parts of the original building such as the dining hall, where large glazed areas and glass bricks in the roof resulted in heat loss, were given a significantly better energy balance by effective measures at detail level. The same applies to the original sliding doors of the gymnasium, which are still in existence and which are used to open the hall fully on one of its longer sides. Here it was also possible to retain the original fixture, while meeting energy and safety requirements with the installation of new glazing on the inside.

A special feature of this building is the fact that the original choice of materials and the exceptionally high standard of craftsmanship must be considered exemplary by today's standards. Examples are the narrow but from the design point of view extremely important copper roof verge and copper flashings which, although eighty years old, were still suitable for retention following a cleaning process: a prime example of sustainability, as they are capable of lasting one generation further in their life cycle.

III
IMPROVEMENT IN ENERGY EFFICIENCY AND QUALITY IN THE SIEDLUNG AM SCHILLERPARK RESIDENTIAL DEVELOPMENT IN BERLIN-WEDDING INTEGRATED PROJECT TITLED "ENERGY AND HISTORIC MONUMENTS POST-WAR ARCHITECTURAL MODERNISM"

From 1955 to 1959, Hans Hoffmann built five rows of four-story, flat-roofed houses between Corker Straße, Dubliner Straße and Holländer Straße in Berlin's district of Wedding for a housing cooperative society, Berliner Bau- und Wohnungsgenossenschaft von 1892. Together with the initial development by Bruno Taut dating from 1924 to 1930, which since 2008 has become a UNESCO world cultural heritage site, they form the landmarked Siedlung am Schillerpark estate.

The balconied facades with their floor-to-ceiling, twin-pane, "flowerbox" windows between the main living room and balcony, which are typical of Hoffman, set a new benchmark in social housing during the 1950s. They were intended to increase daylight and comfort levels and create a generous connection between the interior space and natural surroundings. The layout of the two-and-a-half-room apartments is highly functional. The radiators are located at the centre of the building and hence are extremely economically placed.

The clearly structured blocks with their balanced proportions reflect the resource-saving approach to the use of materials during the post-war era; characteristic of this is the deliberately limited range of facing materials. Although—after more than fifty years—Hoffmann's buildings show signs of wear, they have no serious defects and thereby demonstrate a high quality of design and execution. In view of current energy conservation standards, the refurbishment has focused primarily on improving the thermal insulation of the building envelope and on upgrading the services installations to modern technical and energy conservation standards. These objectives had to be achieved without losing the architectural qualities of these Hoffmann buildings, which are important examples of post-war modernism in Berlin.

In order to find an adequate and integrated solution for these diverse requirements for physical and structural improvements on the one hand, and preservation of the historic building substance on the other, these Hoffmann buildings were selected as a model for the research project "Historic Buildings and Energy Post-war Modernism," initiated with the help of funding from the DBU (Deutsche Bundesstiftung Umwelt).[11]

In order to establish what refurbishment work would be required, we carried out a detailed survey of the building.

9   Stairway to the gallery of the sports hall: The windows of the fenestration band can be opened outwards like the wings of a butterfly.

10   The copper roof flashing on the student residence, as seen prior to refurbishment.

11   After refurbishment: The flashing had been installed with such a high quality of workmanship that it was possible to retain the original material, although it was eighty years old; it only needed mechanical cleaning.

With the support of our partners from the research project, we were able to analyze the thermal properties of the existing building materials carefully, to use thermography for the detection of weak areas from the energy conservation point of view, and to use computer simulation methods to determine comfort levels and air-flow patterns in the rooms, taking into account window designs and the position of radiators. This revealed that the location of the radiators deep inside the building has a very positive effect and that the heat loss from the buildings is primarily due to the poor insulation of the hollow core blocks in the external walls, the cantilevered balcony floor slabs and the window surfaces, as well as outdated heating installations.

Converting the heating system to district heating and replacing radiators and heating pipes resulted in significant energy savings. For the insulation of the facades, in keeping with the landmark building requirement, we selected a sandwich insulation system with phenolic resin foam, which achieves good insulation values using relatively thin layers of material. In contrast to the conventional though significantly cheaper polystyrene, which is extremely problematic when exposed to heat, this material does not pose any problems in the case of fire. For other areas of the facade, for example the gable ends, the use of thicker and more economic insulation materials can be considered. A useful tool for the assessment of such alternative solutions is the LEGEP software program, which can be used to evaluate parameters of commercial viability together with sustainability.

With a mineral-based traditional render coating it is possible to produce a surface finish that is very similar to the original, but with a significantly longer service life. By contrast, a thinner and less costly render coating based on synthetic resin is prone to attack by algae and fungi once the biocide preservatives have leached out, which has a negative effect on the life cycle of this material. When it comes to demolition, the thin application of the render means that the insulation and the render coat would have to be disposed of as hazardous waste.

From the point of view of the preservation of historic buildings, the floor-to-ceiling, twin-pane, "flowerbox window" is the characteristic architectural element of the Hoffmann buildings. We considered the idea of increasing the floor area by removing the inner glazing, as suggested by the client on several occasions, but felt that it would cause problems from the building physics point of view and would not fulfill the preservation criteria. In addition there is the fact that such conversion work is much harder to carry out when the buildings are occupied.

The "flowerbox window" acts as a buffer zone and is very useful for energy conservation, as it considerably reduces heat loss via the thermal bridge of the balcony slab. The large space between the window panes has a positive effect on the interior climate, as it supports the mechanical ventilation concept thus: Air extraction via the chimneys causes negative pressure in the apartment, which in turn causes fresh outside air to be drawn in via ventilation slots in the "flowerbox windows." This incoming air heats up naturally in the space between the window panes.

Another characteristic design feature is the glazing in the stairwells. The intention was to upgrade the thermal insulation properties of the existing steel structure with its very slender profiles by installing secondary double glazing. This would have involved additional interior reinforcement of the framing. An alternative would have been to rebuild the units completely in the form of a thermally separated, aluminum frame construction; however, this meant much wider profiles with a different geometry, which would have lost much of the transparent effect of the existing glazing.

This project is currently in the design phase, with construction scheduled for 2011; therefore we do not yet have any measured energy conservation data. It is our intention here to discuss the broader design concept and to consider its variables as a means of optimizing a solution that strikes a balance between cost and environmental impact.

Over an assessment period of twenty five years, the various options do not show significant differences in terms of potential environmental pollution, as this is primarily affected by the ongoing use of the building and only to a small degree by the manufacturing and maintenance costs.

When considering the energy conservation measures from the point of view of the initial investment, a rather startling fact emerges with regard to rented accommodation: The savings benefit primarily the purse of the tenants rather than that of the owner or investor. Having said that, any energy conservation improvement ultimately benefits the environment and hence all of us.

I   It was fortuitous that we had the opportunity to select the design team ourselves. With Transsolar for building services, Halfkann u. Kirchner as fire safety experts and Pichler Ingenieure for structural engineering and building physics we were able to field our ideal team, both for the competition and the subsequent execution.

II   The objective of the project is to compile, evaluate and document the results of research in building construction for a large number of comparable residential developments. Scientific support for the project is provided by the research group under Prof. Dr. Weller at the Technical University of Dresden, while the multidisciplinary design team is headed by Winfried Brenne Architects. In addition, the involvement of a small window-making firm with experience in historic building work is intended to facilitate the creation of innovative craft-based product solutions.

12   [p. 186] The facade of the Am Schillerpark development features an unusual combination of loggias and balconies. At the same time, the glazing of the rooms aseems to extend all the way to the balconies.

13   The flower window, a characteristic element of housing designed by Hoffmann, functions as a thermal buffer zone and improves the room climate.

GENERAL LITERATURE AND MONOGRAPHS

Bone-Winkel, Stephan: "Projektentwicklung im Bestand". In: Architekten- und Stadtplanerkammer Hessen (ed.): *Planen im Bestand. Bauen für die Zukunft.* Wiesbaden 2005. pp. 58–75.

Breitling, Stefan/Cramer, Johannes: *Architektur im Bestand. Planung, Entwurf, Ausführung.* Basel 2007.

von Buttlar, Adrian/Heuter, Christoph (eds.): *Denkmal Moderne: Architektur der 60er Jahre – Wiederentdeckung einer Epoche.* Berlin 2007.

Dal Co, Francesco/Mazzariol, Giuseppe: *Carlo Scarpa: The Complete Works.* 2002.

Drexel, Thomas: *Faszination Bauernhaus: Renovieren—Umbauen—Erweitern.* Munich 2009.

Ebbert, Thiemo: *Re-Face. Refurbishment Strategies for the Technical Improvement of Office Façades.* TU Delft Architecture, Dissertation 2010.

Gebhard, Helmut/Sauerländer, Willibald (eds.): *Feindbild Geschichte. Positionen der Architektur und Kunst im 20. Jahrhundert.* Göttingen 2007.

Grube, Hans Achim (ed.): *New Power, Elektropolis im Wandel.* Berlin 2006.

Harlfinger, Thomas/Richter, Dirk: "Objektentwicklung von Bestandsimmobilien. Potenzialbestimmende Faktoren." In: *LACER 9/2004.* pp. 77–84.

Heinemann, Andrea/Zieher, Heike: *Bunker update. Vorschläge zum heutigen Umgang mit Bunkern in innerstädtischen Lagen.* Dortmund 2008.

Jester, Katharina/Schneider, Enno: *Weiterbauen.* Berlin 2002.

Klanten, Robert/Feireiss, Lukas (eds.): *Build-On. Converted Architecture and Transformed Buildings.* Berlin 2009.

Klostermeier, Collin/Wieckhorst, Thomas: *Umbauen, Sanieren, Restaurieren. 28 Gebäude aus 8 Jahrhunderten.* Gütersloh 2006.

Los, Sergio: *Scarpa,* Cologne 2009.

Portrait Hans Döllgast, in: Nerdinger, Winfried (ed.): *Süddeutsche Bautradition im 20. Jahrhundert. Architekten der Bayerischen Akademie der Schönen Künste,* exhibition catalog. Munich 1985. pp. 251–290.

Noever, Peter (ed.): *Carlo Scarpa. Das Handwerk der Architektur/The Craft of Architecture,* MAK exhibition catalog. Ostfildern-Ruit 2003.

O'Kelly, Emma/Dean, Corina: *Conversions.* London 2007.

Pehnt, Wolfgang: *Karljosef Schattner: Ein Architekt aus Eichstätt.* Ostfildern 1988/1999.

Pehnt, Wolfgang: "Amnesie statt Anamnese. Über Rekonstruktion, Reproduktion, Remakes und Retro-Kultur." In: *DAM Jahrbuch 2004. Architektur in Deutschland.* Munich 2004.

Pehnt, Wolfgang: "Dem Bau zu sich selbst verhelfen. Burg Rothenfels und die interpretierende Denkmalpflege." In: Ingrid Scheuermann (ed.): *ZeitSchichten. Erkennen und Erhalten—Denkmalpflege in Deutschland.* Catalog Residenzschloss Dresden. Munich/Berlin 2005. pp. 124 ff.

Powell, Kenneth: *Architecture Reborn: The Conversion and Reconstruction of Old Buildings (Masterpieces of Architecture).* London 2005.

Prudon, Theodore H. M.: *Preservation of Modern Architecture.* Hoboken 2008.

Ringel, Johannes/Bohn, Thomas/Harlfinger, Thomas: "Objektentwicklung im Bestand—aktive Stadtentwicklung und Potentiale für die Immobilienwirtschaft!?" In: *Zeitschrift für Immobilienwirtschaft 1/2004.* Cologne 2004. pp. 45–52.

Santifaller, Enrico (ed.): *Transform. Zur Revitalisierung von Immobilien, The Revitalisation of Buildings.* Munich/Berlin/London/New York 2008.

Schattner, Karljosef: *Karljosef Schattner: Ein Führer zu seinen Bauten.* Munich 1998.

Schittich, Christian: *Bauen im Bestand. Umnutzung, Ergänzung, Neuschöpfung.* Basel 2003.

Thiébaut, Pierre: *Old buildings looking for new use. 61 examples of regional architecture between tradition and modernity.* Stuttgart/London 2007.

TU Munich (ed.): *Hans Döllgast 1891–1974,* exhibition catalog. Munich 1987.

Waiz, Susanne: *Auf Gebautem bauen. Im Dialog mit historischer Bausubstanz. Eine Recherche in Südtirol.* Vienna/Bolzano 2005.

Wehdorn, Jessica: *Kirchenbauten profan genutzt. Der Baubestand in Österreich.* Vienna/Bolzano 2006.

Weidinger, Hans: *Einfamilienhäuser von 1960–1980 modernisieren. Renovieren–Anbauen–Umbauen–Aufstocken.* Munich 2003.

Wüstenrot Stiftung (ed.): *Umbau im Bestand.* Stuttgart/Zurich 2008.

PERIODICALS

B+B. Bauen im Bestand. Cologne.

Internationale Zeitschrift für Bauinstandsetzen und Baudenkmalpflege. Freiburg.

Metamorphose. Bauen im Bestand. Leinfelden-Echterdingen.

Detail – Review of Architecture + Construction Details. Institut für internationale Architektur-Dokumentation. Munich.

SELECTED PUBLICATIONS FROM AUTHORS IN THIS BOOK

Brenne, Winfried: "Practical Experience with the buildings of the Avant-garde in Berlin and East Germany." In: Haspel, Jörg/Petzet, Michael et al. (eds.): *Heritage at risk—The Soviet Heritage and European Modernism.* Berlin 2007. pp. 146–150.

Brenne, Winfried: *Bruno Taut—Meister des farbigen Bauens in Berlin,* ed. by Deutscher Werkbund Berlin. Berlin 2005/2007.

Brenne, Winfried: "Work experience with buildings of the modern movement." In: Kudryavtsev, Alexander (ed.): *Heritage at risk—Preservation of the 20th century architecture and world heritage.* Moscow 2006. pp. 43–44.

Brenne, Winfried: "Die Revitalisierung eines Denkmals." In: *Das Berliner Zeughaus—vom Waffenarsenal zum Deutschen historischen Museum,* ed. by Ulrike Kretzschmar. Munich, Berlin, London, New York 2006. pp. 98–105.

Brenne, Winfried: "Instandsetzungsplanung zwischen Erhaltung, Reparatur und Neubau." In: Wüstenrot Stiftung Ludwigsburg (ed.): *Meisterhaus Muche/Schlemmer—Die Geschichte einer Instandsetzung.* Stuttgart 2003. pp. 111–133.

Brenne, Winfried: "L'edilizia residenziale di Bruno Taut. Conservazione e recupero dell'architettura del colore." In: Nerdinger, Winfried/Speidel, Manfred/Hartmann, Kristiana/Schirren, Matthias (eds.): *Bruno Taut 1880–1938.* Milan 2001. pp. 275–289.

Brenne, Winfried: "Die 'farbige Stadt' und die farbige Siedlung—Siedlungen von Bruno Taut und Otto Rudolf Salvisberg in Deutschland." In: *Mineralfarben— Beiträge zur Geschichte und Restaurierung von Fassadenmalereien und Anstrichen.* publication of the Institute for Historic Building Research and Conservation at the Swiss Federal Institute of Technology (ETH) Zurich, volume 19. Zurich 1998. pp. 67–78.

Brenne, Winfried/Pitz, Helge: *Siedlung Onkel Tom—Einfamilienreihenhäuser 1929. Die Bauwerke und Kunstdenkmäler von Berlin,* supplement 1, ed. by Senator for Building and Housing, State Conservator. Berlin 1980 (new edition 1998).

Brenne, Winfried: "Die intelligente Farbe – Mit Farbe bauen." In: Speidel, Manfred (ed.): *Bruno Taut —Natur und Fantasie 1880–1938.* Berlin 1995. pp. 228–231.

Hempel, Rainer: "Historische Tragwerke: Zisterzienserkloster Walkenried." In: Fakultät für Architektur der Fachhochschule Köln (ed.): *Baudenkmalpflege in Lehre und Forschung, Festschrift für Prof. Dr.-Ing. J. Eberhardt der FH Köln.* Cologne 2003. pp. 42–48.

Hempel, Rainer: "Statisch-konstruktiver Brandschutz im Bestand." In: *VdS-Fachtagung Brandschutz, Grenzen des Brandschutzes, Tagungsband VdS.* Cologne 2004.

Hempel, Rainer: "Projekt Revitalisierung Verwaltungsgebäude Dorma GmbH & Co. KG Ennepetal" (in cooperation with KSP Engel u. Zimmermann architects, Cologne). In: *Der Baumeister 10/2004, Stahlbau 09/2004; Bauen mit Stahl 2005; Die Neuen Architekturführer Nr. 65.* Berlin 2005.

Jäger, Frank Peter: *Neues Quartier Vulkan Köln—Leben und Arbeiten im Industriedenkmal.* Berlin 2007.

Jaschke, Franz/Brenne, Winfried: "Die Sanierung von Siedlungsbauten der klassischen Moderne—Langzeiterfahrung und Know-how eines Berliner Architekturbüros." In: *Die Siedlung Freie Scholle in Trebbin,* ed. by Raimund Fein/Markus Otto/Lars Scharnholz. Cottbus 2002. pp. 37–43

Pottgiesser, Uta: *Fassadenschichtungen—Glas. Mehrschalige Glaskonstruktionen: Typologie, Energie, Konstruktionen, Projektbeispiele.* Berlin 2004.

Pottgiesser, Uta: "Revitalisation Strategies for Modern Glass Facades of the 20th century." In: *Proceedings STREMAH 2009. Eleventh International Conference on Structural Studies, Repairs and Maintenance of Heritage Architecture.* Southampton 2009.

Rexroth, Susanne: "Photovoltaik im historischen Bestand." In: *Gebäudeintegrierte Photovoltaik,* conference proceedings from OTTI orientation seminar, March 2009. pp. 99–104.

Rexroth, Susanne: "Alte Hülle—zeitgemäße Energiebilanz." In: *archplus 184,* 10/2007. pp. 98 f.

Rexroth, Susanne: "Atmosphären—subjektiv und objektiv: Maßnahmen und Techniken zur Energieeinsparung an Baudenkmalen." In: *Energieeffiziente Sanierung von Baudenkmalen und Nichtwohngebäuden,* conference proceedings. Institute for Building Construction at TU Dresden. Dresden 2007. pp. 45–52.

SELECTED LITERATURE ON THE DEPICTED PROJECTS AND ARCHITECTURAL OFFICES

Mineralogical Collection in Castle Freudenstein/AFF Architekten:

AFF Architekten: *Teile zum Ganzen/An Aggregate Body. Schloss Freudenstein.* Tübingen/Berlin 2009.

Würzburg Cogeneration Plant/Brückner & Brückner:

Santifaller, Enrico (ed.): *Stadtraum und Energie. Heizkraftwerk Würzburg.* Passau 2009.

Chesa Albertini/Hans-Jörg Ruch Architektur:

Ruch, Hans-Jörg: *Historische Häuser im Engadin. Architektonische Interventionen.* Zurich 2009.

Blumen Primary School and Bernhard Rose School/huber staudt architects bda:

Detail 9/2009, pp. 894–901.

Metamorphose. Bauen im Bestand, 03/08, pp. 52 ff.

Cafeteria in the Zeughaus Ruin/Kassel Building Department, Prof. Hans-Joachim Neukäter:

Krüger, Boris/Müller, Volker: *Das Zeughaus in Kassel. Bilder aus seiner Geschichte.* Kassel 2004.

Pier Arts Centre/Reiach and Hall Architects:

Reiach and Hall Architects: *The Pier Arts Centre Stromness Orkney.* Edinburgh 2007.

Public Library in Ebracher Hof/Bruno Fioretti Marquez Architekten:

Kleefisch-Jobs, Ursula: "Stadtbücherei im Ebracher Hof." In: Peter Cachola Schmal (ed.): *Deutsches Architektur Jahrbuch, German architecture annual 2008/2009.* Munich 2009.

Rose am Lend/Innocad Architekten:

Ruby, Andreas/Ruby, Ilka (eds.): *Von Menschen und Häusern. Architektur aus der Steiermark. Graz-Styria architectural yearbook 2008/2009.* Graz 2009.

ARCHITECTS' WEBSITES

Interview:
www.meixner-schlueter-wendt.de
www.architektenbrueckner.de
Addition:
www.architektenbrueckner.de
www.reiachandhall.co.uk
www.sunder-plassmann.com
www.peterkulka.de
www.search-arch.ch
www.luederwaldt-architekten.de
www.innocad.at
www.rkw-as.de
www.numrich-albrecht.de
www.architektenbrueckner.de
Transformation:
www.ruch-arch.ch
www.andotadao.org
www.broadwaymalyan.com
www.bolwinwulf.de
www.coastoffice.de
www.adrianstreich.ch
www.karo-architekten.de
www.hhf.ch
www.huberstaudtarchitekten.de
www.heinlewischerpartner.de
www.numrich-albrecht.de
www.schneider-schumacher.de
Conversion:
www.kokoarch.com
www.op-architekten.com
www.anderhalten.com
www.lin-a.com
www.duncanmccauley.com
www.merkx-girod.nl
www.bfm-architekten.de
www.aff-architekten.com

PHOTO CREDITS

p.7	Walter Nauerschnig, Berlin [1]
p.8	Haydar Koyupinar, Bayerische Staats-gemäldesammlungen [2]; Jens Willebrand, Cologne [3, 4]; Jeroen Musch, Rotterdam [5]
p.9	Stiftung Preußischer Kulturbesitz/David Chipperfield Architects, Jörg von Bruchhausen, Berlin [6]; Andrea Wandel, Wandel Höfer Lorch Architekten + Stadtplaner, Saarbrücken [7]; Monika Marasz, Detmold [8]
pp.12,13	Christoph Kraneburg, Cologne [1, 2, 3]
p.15	Constantin Meyer, Cologne
p.17	Heizkraftwerk Würzburg GmbH [1]; Pier Arts Centre [2]; Verein Zeughaus Kassel e.V./Werner Lengemann [3]; Museum Kunst der Westküste, Alkersum [4]; SeARCH [6]; Innocad Architekten, Graz [8]; RKW Architektur + Städtebau, Leipzig [9]
pp.18–20	Constantin Meyer, Cologne
p.22	Reiach and Hall Architects
p.23	Gavin Fraser/ FOTO-MA [2]; Ioana Marinescu [3, 4]
p.24	Alistair Peebles, Pier Arts Centre
p.25	Alistair Peebles, Pier Arts Centre [6]; Gavin Fraser/FOTO-MA [7]; Reiach and Hall Architects [8]; Ioana Marinescu [9]
p.26	Reiach and Hall Architects
p.27	Ioana Marinescu [10, 12]; Alistair Peebles, Pier Arts Centre [11]
pp.28–30	Constantin Meyer, Cologne
p.29	Max-Eyth-Schule/Verein Zeughaus Kassel e.V./Werner Lengemann [2]; Christian Lemke, Kassel [3]
pp.32–37	Frank Grießhammer, The Hague
pp.38–40	Jörg Schöner, Dresden
p.41	Peter Kulka Architektur
pp.42–44	Christian Richters
p.45	SeARCH
pp.46,48,49	Lukas Roth, Cologne
p.47	Lüderwaldt Architekten
pp.50,52	© paul ott photografiert
p.51	Innocad Architekten, Graz
pp.54,55	Industrieterrains Düsseldorf Reisholz AG/Dr. Pröpper [1, 2, 3]
pp.55,56	RKW Architektur + Städtebau [4, 5, 6, 7]
p.57	RKW Architektur + Städtebau [8, 9, 10]; Gunter Binsack, Leipzig [11]
pp.58–61	Gunter Binsack, Leipzig
pp.62–65	Filippo Simonetti, Brunate
p.67	huber staudt architekten bda, Berlin [9]; tadao ando architect & associates [1]; Broadway Malyan [2]; COAST office architecture [4]; KARO* Architekten, Leipzig [6]; Peter Affentranger Architekt [8]; Adrian Streich Architekten [5]; Bolwin Wulf Architekten, Berlin [3]; Gottfried Planck, Universitätsbauamt Stuttgart and Hohenheim [10]; HICOG (High Commissioner Germany) [12]; Numrich Albrecht Klumpp Architekten/IGZ Großbeeren [11]; HHF Architekten [7]
pp.68–71	Andrea Jemolo [1, 2, 3, 4, 5]; Alessandra Chemollo, Palazzo Grassi [6]
pp.72,73,75	Fernando Guerra, Lissabon
p.74	Broadway Malyan
pp.76–79	Rolf Sturm, Landshut
pp.80–83	David Franck Photographie
pp.84–86	Roger Frei, Zurich
p.85	Adrian Streich Architekten [3]
pp.88–90	Anja Schlamann, Cologne/Leipzig
p.90	KARO* Architekten [3]
pp.92–94	Tom Bisig, Basel
p.93	HHF Architekten
pp.95–97	Peter Affentranger Architekt
pp.98–100	Werner Huthmacher, Berlin
pp.102–107	Werner Huthmacher, Berlin [1]; huber staudt architekten bda, Berlin [2, 3]; Michael van Ooyen, Straelen [4]; Klaus Legner, Moers [5, 9]; Jens Willebrand, Cologne [6, 7]; Brigida González [8]
pp.108–110	Brigida González
p.109	Jogi Hild Fotografie [4]
p.111	Heinle, Wischer und Partner
pp.112–115	Nina Straßgütl
p.114	Numrich Albrecht Klumpp Architekten
pp.116,118	Jörg Hempel Photodesign, Aachen
p.117	HICOG (High Commissioner Germany)
p.119	Martin Brück 2009, Stiftung Bauhaus Dessau [1]; Uta Pottgiesser [2]
p.120	Uta Pottgiesser [3]; Wolfgang Nigescher [4]; Stefan Müller, Berlin [5]; Uta Pottgiesser [6]
p.123	Stefan Müller, Berlin [5]; Uta Pottgiesser [6]
p.124	Martin Brück 2009, Stiftung Bauhaus Dessau [1]; Stefan Müller, Berlin [5]
pp.125,126	Kai Oswald Seidler [7]; Julia Jungfer, Berlin [10]; Frank Peter Jäger [11]
p.126	Uta Pottgiesser [8]; Tchoban Voss Architekten [9]
pp.127,128	H. G. Esch, Hennef [12, 13]
pp.128,129	HPP Architekten [14]; Uta Pottgiesser [15, 16]
p.131	Ursula Böhmer, Berlin [4]; Kaido Haagen [1]; Merkx + Girod Architecten [7]; LIN Finn Geipel + Giulia Andi [5]; Bruno Fioretti Marquez Architekte [9]; Numrich Albrecht Klumpp Architekten [3]; OP Architekten [2]; Brückner & Brückner Architekten [8]; Duncan McCauley [6]
pp.132,133	Arne Maasik [1, 2, 3]; Kaido Haagen [4]
p.135	Fred Laur [5, 8, 9]; Kaido Haagen [6]; Vallo Kruser [7]
p.136	Ula Tarasiewicz, OP Architekten
pp.138,139	Wojciech Popławski [2]; Wallphotex, OP Architekten [3, 4, 5, 6, 7]
pp.143–144	Hempel & Partner Ingenieure, Cologne
pp.146–147	Hempel & Partner Ingenieure, Cologne
pp.145,146	Architekten Kleineberg & Pohl, Braun-schweig [section/plan]
pp.148,149,151	Werner Huthmacher, Berlin
pp.150	Numrich Albrecht Klumpp Architekten
pp.152,153	Ursula Böhmer, Berlin
pp.154,155	Werner Huthmacher, Berlin [2, 3, 4, 6]; Ursula Böhmer, Berlin [5]
pp.156,157	Christian Richters [1, 4]; Hans-Michael Földeak, LIN [2]; Jan-Oliver Kunze, LIN [3]
p.159	Jan-Oliver Kunze, LIN [5]; Christian Richters [6, 7]
p.160	Christian Richters [8, 9, 10]
pp.161–163	Jan Bitter, Berlin
pp.164,166,167	Roos Aldershoff Fotografie
pp.168,169,171	Peter Manev, Selb
pp.172,174–177	Annette Kisling, Berlin [1, 2, 3, 7, 9]; Christoph Rokitta, Berlin [4, 6, 10]; City of Schweinfurt [5]
pp.178,180,181	Sven Fröhlich, AFF Architekten
pp.183–188	Holger Herschel, Berlin [1, 2, 4, 5, 6, 7, 8]; Winfried Brenne Architekten [3, 9, 10, 11, 12, 13]

EDITOR
FRANK PETER JÄGER

Frank Peter Jäger's fascination for cities, urban land-scapes, and their buildings reaches back to his child-hood. His earliest, formative impressions were of the row houses and countless multistory buildings of a monotonous 1960s housing development, as well as the monumental brick shells of the briquette factories and power plants of the brown coal territory along the Rhine; and in addition—pale blue in the back-ground—the Gothic mountain of Cologne's Cathedral. The desire to combine worthwhile buildings from the past with those of the future emerged early on, and ultimately led to the topic of working within the ex-isting fabric.

Dipl.-Ing. Frank Peter Jäger works as a publicist and public relations consultant for architects; he is also an instructor at universities and gives continuing educa-tion courses at institutions that include local cham-bers of architects. He received his journalistic training in the vastness of Brandenburg and at the newspaper *FAZ*; since then, he works as an author and architec-ture critic for various daily newspapers and trade media. Simultaneously, he has produced a number of books, which include *Dorotheenhöfe*, a coffee-table book of Oswald Mathias Ungers' works in Berlin. A familiar subject for Jäger is the professional practice of architects and planners: his PR guide for architects, *Offensive Architektur* appeared in 2004, and in 2008 he edited *Der neue Architekt—Erfolgreich am verän-derten Markt*. Frank Peter Jäger lives in Berlin and has a six-year-old son.

www.archikontext.de/jaeger@archikontext.de

AUTHORS OF THE PROJECT PORTRAITS
HUBERTUS ADAM

Hubertus Adam was born in 1965 in Hanover and studied art history, philosophy, and archaeology in Heidelberg. Since 1992, he has worked as a freelance historian of art and architecture and as an architec-tural critic for various magazines and newspapers, in-cluding the *Neue Zürcher Zeitung*. In 1998 he moved to Switzerland, where he edits the magazine *archithe-se*. In addition, he serves as jury member, presenter, and curator. Numerous books and book contributions. Swiss Art Award in 2004 for the section on art and architectural education.

ANNEKE BOKERN

Anneke Bokern was born in 1971 and studied art his-tory in Berlin. Since 2000 she has lived in Amsterdam, where she writes on architecture and design as a free-lance journalist. Her articles have appeared in *Bauwelt, Baumeister, db, Metamorphose, design report, NZZ, Frame, Mark* and *DAMn° Magazine*.
www.anneke-bokern.de

CLAUDIA HILDNER

Claudia Hildner studied architecture in Munich and Tokyo. Subsequently served a traineeship with the magazine Baumeister; since 2007, she has worked as a freelance journalist and editor for various architec-tural magazines and publishers. Her areas of focus are building within the existing fabric, and Japan.
www.childner.de

SIMONE JUNG

Simone Jung was born in 1978 in Hesse. After com-mercial vocational training, she studied journalism, sociology, and art history in Mainz. Her practical ex-perience has been gathered in the culture and media sector (incl. *N24, Deutsche Welle, De-Bug, Frankfurter Allgemeine Sonntagszeitung, taz*). She now lives in Berlin and writes about culture and society as a free-lance author.

FRANK VETTEL

Frank Vettel was born in 1964 and studied architec-ture and urban design at TU Darmstadt. Since 1992, he has worked in various architectural offices in Berlin. In 1998, he passed the second state board examina-tion after internship in building construction. Since 1998 he has been employed by the borough of Berlin-Friedrichshain, where in 2003 he became director of building construction, and in 2007 director of facility management. Competition entries include the Berlin Museum (with Stefan Forster).

TEXT SOURCES

The descriptions of the projects are by the editor (FPJ) or the authors named at the beginning of the respective text. The texts for the following projects are based on descriptions written by the correspond-ing architectural offices, which have been revised and amended by the editor:

[p. 28]   Cafeteria in the Zeughaus Ruin,
          Kassel Building Dept./ Hans-Joachim
          Neukäter
[p. 38]   Roof for the Kleiner Schlosshof,
          Peter Kulka Architektur
[p. 76]   Sparkasse Berchtesgadener Land,
          Bolwin Wulf Architekten
[p. 80]   Weinstadt Town Hall, COAST office
          architecture
[p. 84]   Heuried Residential Complex,
          Adrian Streich Architekten AG
[p. 95]   School in Dagmersellen,
          Peter Affentranger Architekt
[p. 116] Siesmayerstraße Office Building,
          schneider+schumacher
[p. 152] Wildau Laboratory Building,
          Anderhalten Architekten
[p. 156] Alvéole 14 Cultural Center,
          LIN Finn Geipel + Giulia Andi

ACKNOWLEDGMENTS

No architectural book can be accomplished without the energetic support of those who have designed and built the projects. Therefore many thanks to the municipal building department of Kassel and to Prof. Hans-Joachim Neukäter—as well as to all the archi-tectural offices whose buildings are represented in the book—for providing, and in some cases preparing, the project material.

The receptiveness of Claudia Meixner, Florian Schlüter, and the brothers Peter and Christian Brückner to my idea for the interview pleased me very much.

All twelve co-authors of this book have invested much enthusiasm and energy in their contributions. Thanks to all, and especially to Prof. Dr. Rainer Hempel, for preparing the plans and sections of the projects to meet our layout needs. Prof. Dr. Uta Pottgiesser and Julia Kirch of the University of Applied Sciences in Ostwestfalen-Lippe have gone well beyond what could be expected in providing assistance for Old & New, including the "Lisbon research" from their colleague Luiza Corrêa. Simone Jung and Teodora Vasileva have rendered outstanding services by con-ducting research and by procuring materials, which was not without its complications.